Biotechnology in Tall Fescue Improvement

Editor

M. J. Kasperbauer

Research Plant Physiologist
Coastal Plain Research Center
U.S. Department of Agriculture
Agricultural Research Service
Florence, South Carolina

CRC Press
Boca Raton Ann Arbor Boston

Library of Congress Cataloging-in-Publication Data

Biotechnology in tall fescue improvement / editor, M.J. Kasperbauer.
p. cm.
Includes bibliographical references.
Contents: Importance and problems of tall fescue / Robert F. Barnes -- Potential biotechnological approaches / Karan Kaul -- Somatic cell culture and embryogenesis in the Poaceae / D.J. Gray -- Plant regeneration and evaluation / M.J. Kasperbauer -- Haploids / M.J. Kasperbauer -- Doubled haploids / M.J. Kasperbauer -- Cytology of haploids and doubled haploids / Warren D. Springer / Physiology and biochemistry / C.J. Nelson and D.A. Sleper -- Breeding and genetics / D.A. Sleper and C.J. Nelson.
ISBN 0-8493-4891-9
1. Tall fescue--Biotechnology. I. Kasperbauer, M. J., 1929-
SB201.T34B56 1990
633.2'8--dc20 89-25379
CIP

Direct all inquiries to CRC Press, Inc., 2000 Corporate Blvd., N.W., Boca Raton, Florida 33431.

International Standard Book Number 0-8493-4891-9

Library of Congress Card Number 89-25379
Printed in the United States

PREFACE

The idea that eventually led to this book originated during conversations with traditional plant breeders and with scientists working at the cellular and subcellular levels. It became apparent that traditional and new technologies should complement each other in order to be most useful in developing superior new genetic lines, hybrids, or cultivars. It was also apparent that the first step in such an approach should be to identify real problems that are difficult or extremely slow to solve by traditional methods. The next step should then be to consider the various new technologies in terms of their potential to solve the problems.

This book begins with a discussion of the widespread importance of tall fescue as a forage and turf grass, and the difficulties encountered in genetic improvement of this species because it is wind pollinated and generally self-infertile. The pros and cons of various biotechnological approaches are discussed. Because tall fescue is highly heterozygous, the book concentrates on using the natural genetic variability through culture of haploid plants from immature gametes, selection to identify superior haploid plants, and regeneration of fertile doubled haploids with the genetic characteristics identified in the selected haploids. The last two chapters are by a physiologist/breeder team who discuss the potential of haploids and doubled haploids in conjunction with conventional approaches for more rapid genetic improvement of physiological processes that contribute to improved forage and turf characteristics. Throughout the book, chapter authors reflect on problems to be solved and the rationale behind the various approaches.

M.J.K.
May 1989

THE EDITOR

Michael John Kasperbauer, Ph.D., is Research Plant Physiologist in the Agricultural Research Service, U.S. Department of Agriculture at Florence, South Carolina, and Professor (Adjunct) of Plant Physiology at Clemson University.

Dr. Kasperbauer received his Ph.D. degree in 1961 from Iowa State University. He was awarded a National Science Foundation Postdoctoral Fellowship at the University of Maryland in cooperation with the USDA Pioneering Research Laboratory for Plant Physiology in 1961 to 1962. This was followed by a year as the National Academy of Science/National Research Council Resident Research Associate at the USDA Pioneering Research Laboratory for Plant Physiology at Beltsville, Maryland in 1962 to 1963. He was appointed Research Plant Physiologist in the USDA, Agricultural Research Service Tobacco and Forage Research Unit at the University of Kentucky from 1963 to 1983. He became Associate Professor (Adjunct) and member of the Graduate Faculty in 1965 and Professor (Adjunct) in 1973. He transferred to the USDA-ARS Coastal Plains Research Center at Florence, South Carolina in 1983, and was also appointed Professor (Adjunct) of Plant Physiology at Clemson University.

Dr. Kasperbauer is a member of the American Society of Plant Physiologists, Scandanavian Society for Plant Physiologists, American Society of Agronomy, Crop Science Society of America, Southern Association of Agricultural Scientists, International Congress of Photobiology, International Botanical Congress, International Grassland Congress, and the International Congress of Plant Tissue and Cell Culture. He has been elected to honor societies: Phi Kappa Phi, Gamma Sigma Delta, Phi Delta Kappa, and Sigma Xi.

Dr. Kasperbauer has authored or coauthored more than 150 research papers. He served as Associate Editor of Agronomy Journal from 1975 to 1983. He was elected Fellow of the Crop Science Society of America and Fellow of the American Society of Agronomy.

CONTRIBUTORS

Robert F. Barnes, Ph.D.
Executive Vice President
American Society of Agronomy
Crop Science Society of America
Soil Science Society of America
Madison, Wisconsin

Dennis J. Gray, Ph.D.
Associate Professor and Developmental Biologist
Central Florida Research and Education Center
University of Florida
Leesburg, Florida

Michael J. Kasperbauer, Ph.D.
Research Plant Physiologist
Coastal Plains Research Center
Agricultural Research Service
United States Department of Agriculture
Florence, South Carolina

Karan Kaul, Ph.D.
Coordinator
Plant and Soil Science Research
Kentucky State University
Frankfort, Kentucky

C. J. Nelson, Ph.D.
Curators Professor
Department of Agronomy
University of Missouri
Columbia, Missouri

David A. Sleper, Ph.D.
Professor
Department of Agronomy
University of Missouri
Columbia, Missouri

Warren D. Springer, Ph.D.
Department Head
Department of Tissue and Cell Biology
Northrup King Company
Stanton, Minnesota
Currently:
Manager, Regulatory Affairs
Northrup King Company
Minneapolis, Minnesota

TABLE OF CONTENTS

Chapter 1

IMPORTANCE AND PROBLEMS OF TALL FESCUE

Robert F. Barnes

TABLE OF CONTENTS

I. INTRODUCTION

Tall fescue (*Festuca arundinancea* Schreb.) is a wind-pollinated, highly self-infertile polyploid perennial cool-season forage and turfgrass. It is used in pastures, parks, lawns, golf courses, football fields, highway medians and roadsides. Tall fescue serves as perennial ground cover for millions of acres of erodible land. Thus, it is important for grazing, stabilizing soil for agriculture, and enhancing the environment through multiple uses, such as forage, conservation, and turf.[1,2]

Tall fescue has received exceptional visibility and acceptance since the 1940s. It has attained widespread usage as a forage grass for livestock, despite its propensity to cause physiological disorders, at times, in grazing livestock. Considerable progress has been made in identification of causes of the disorders associated with grazing of fescue.

With the discovery of its infection by an endophytic fungus (*Acremonium coenophialum*), a new thrust of interest has been generated in tall fescue. Promise for continued multiple uses in the future is great, but many problems remain to be resolved. As with many forage species, its success is dependent upon improved germplasm, new cultivars, and better management practices. Improvement through conventional forage grass breeding is a slow process. However, new research tools, such as tissue and anther culture, offer a unique approach to make more rapid improvements in an already valuable grass.

II. DISTRIBUTION AND ADAPTATION

Tall fescue is indigenous to Europe where it is well adapted. It occurs on the Baltic coasts, throughout the Caucasus, and in western Siberia, extending into China. Introductions have been made into Australia, Canada, Japan, New Zealand, North and South America, and South and East Africa.

Fescue was introduced to North America in the early to mid 1800s. Sources of tall fescue and meadow fescue (*Festuca pratensis* Huds) were tested at various locations in the U.S. during the late 1800s and early 1900s. However, tall fescue received little attention until it was recognized for its value as a forage in the late 1930s.

"Kentucky 31" tall fescue originated from an ecotype found growing in Menifee County, Kentucky, before 1890. It was brought to the attention of Professor E. N. Fergus (University of Kentucky) during 1931. Grass produced from seed collected at the site was tested and subsequently released as a cultivar (Kentucky 31) in 1943.[1] The cultivar, "Alta", was selected from a series of plant introductions on the basis of winter survival in the northwestern part of the U.S. and released in 1940 by the Oregon Agricultural Experiment Station and the U.S. Department of Agriculture, Agricultural Research Service.[3]

Approximately 400,000 acres of tall fescue were grown in 1940 in the U.S. It has since become the predominant cool-season perennial grass species grown on an estimated 35 million acres.[1] Some of the acreage has resulted from natural seeding, but much of the acreage is the result of introduced seedings.

Tall fescue is a leafy, course-textured, vigorous perennial bunchgrass that reproduces vegetatively through tillering and by seed. It has many agronomic and forage attributes such as hardiness and desirable chemical composition. Although relatively easy to establish, tall fescue is slow to establish from seed, due to low seedling vigor, compared to that of orchardgrass and ryegrass. Thus, seeding with companion crops such as small grains, legumes, or other grasses may suppress tall fescue seedling and stand development.

Once established, however, growth is aggressive and competitive, and plants persist under a wide range of environmental conditions. The adaptation and distribution of tall fescue are influenced by climatic (rainfall and temperature), edaphic (soil water and texture), and geographic (latitude and elevation) factors.[4] In the U.S., fescue does best in the transition zone between the cool and warm humid regions, east of the line of 25-in. rainfall in eastern Kansas and Oklahoma, to the Piedmont Region on the Atlantic Coast, and from mid Georgia to southern Ohio, extending as far north as New York.

Plants survive under acid (pH 4.7) to alkaline (pH 9.5) soils and are relatively salt tolerant. Growth occurs on soils ranging from low-lying sites with moist, heavy soils to poorly drained and droughty sites. Fescue is the only reliable perennial cool-season grass on some shallow droughty soils. The coarse, deep root system forms a dense sod which also withstands water logging and flooding. Thus, stands survive for years, if properly managed. The dense, green turf is resistant to animal traffic and can withstand considerable grazing. Excellent turf characteristics make tall fescue particularly valuable for industrial parks, lawns, athletic fields, revegetation of roadsides and disturbed areas, and other recreational or conservation purposes.

Growth starts in early spring, followed by semidormancy during the hot portion of the summer, with growth resuming in the fall and continuing into early winter when other grasses are less productive. Because of abundant autumn growth, forage may be stockpiled beginning in early August. Stockpiling refers to the practice of allowing forage to accumulate in the field until it is needed for grazing.[5] Forage stockpiled for late fall and early winter grazing may be more nutritious than spring-grazed fescue.[6]

III. GENETICS AND BREEDING PROBLEMS

Festuca is a diverse and widely adapted genus which contains over 80 species, of which approximately 20 exist in the U.S.[7] It was not until 1950 that tall fescue was given the name *Festuca arundinacea*.[8] Prior to that date, a taxonomic debate ensued over whether meadow fescue and tall fescue constituted separate distinct species or one broad species.

Terrell[7] noted that further plant exploration is warranted as a means of assessing the precise geographic distribution and natural hybridization among the component taxa of the *Festuca* genus. Such efforts should then be followed by further cytogenetic and taxonomic evaluation.

Cytological analysis of tall fescue is difficult because of the nature of the chromosomes.[9] Although tall fescue is a hexaploid, other ploidy levels and polyhaploids have been reported. Induced endomitotic divisions in tissue culture can be used to produce higher ploidy levels in tall fescue.[10] Such materials have been used to compare the relationship of forage quality characteristics with ploidy level. Increased concentrations of protein, minerals, and digestible nutrients were demonstrated with the higher ploidy levels.[11] More information is needed on the regulatory mechanism controlling chromosome pairing in order to understand the evolutionary sequence in fescues.[9]

The high adaptive potential of tall fescue may be due in part to its polyploidy, in that survivability of chromosome deviation is better in polyploids than diploids.[9] Thus, chromosomes altered through intraspecific hybridization have a good chance to survive because tall fescue is hexaploid, perennial, and can reproduce vegetatively through the tillering process. Similarly, chromosome alteration through tissue culture and other biotechnological approaches, should be considered as part of the fescue improvement program.[10,11]

Transfer of traits from one species to another, through interspecific and intergeneric hybridization, shows promise. The success rate will depend upon the genotype of the parents. Sterility often occurs in the progeny, and the process of transferring favorable genes is likely to be time consuming if one depends entirely on traditional methods of forage grass breeding.[9]

A number of intergeneric hybridization programs involving *Festuca* and *Lolium* have been initiated in the U.S. and several European countries. Progress prior to 1979 was reviewed and summarized in the American Society of Agronomy monograph on tall fescue.[1] In general, intergeneric hybridization allows the recombination of specific characters and increased genetic variation. The procedure has been used to improve winterhardiness, seedling vigor, early spring growth, and forage quality. Two cultivars, ''Kenhy'' and ''Johnstone'', were developed cooperatively by the U.S. Department of Agriculture-Agricultural Research Service and the University of Kentucky utilizing ryegrass-tall fescue hybrid derivatives.[12]

Although genetic studies to determine inheritance of specific characters in tall fescue are limited, rather high heritability estimates have been obtained for some of the traits studied. Characters studied and found to be heritable in tall fescue include seed yield, the alkaloid perloline, net carbon exchange, succulence, leaf width, and mineral content.[1,9,13-15] Tall fescue lines have been shown to exhibit a range of root diameters.[16] The larger diameter roots were associated with improved stand persistence, plant survival, and sward cover. Other traits for which selection should be possible include greenness, seedling vigor, early spring growth, winterhardiness, drought tolerance, forage yield, and disease resistance. However, considerable time and effort would be required by traditional methods to manipulate the genetic code in order to improve a specific genotype and develop a unique cultivar.

Cell culture and plant regeneration techniques may be used to enhance forage grass improvement by allowing selection at the cellular level and regeneration of fertile plants from various selected cells. Methodology has been developed to establish callus cultures from somatic tissue of tall fescue, to modify the cells in culture, and to regenerate numerous plants from the callus,[17] and to culture haploid plants from immature gametes.[18] Kasperbauer and Eizenga[10] developed a tissue culture method to obtain fertile doubled haploid plants with the characteristics identified in the infertile haploid plants. The use of such advanced genetic technology can greatly accelerate the development of improved traits in tall fescue and other forage grasses (see Chapters 3 to 7).

IV. PROBLEMS ASSOCIATED WITH SOIL-PLANT-ANIMAL RELATIONSHIPS

The first step in improving any crop plant is to identify problems. In the case of tall fescue, the primary problems are those related to forage quality, physiological disorders of animals associated with consumption of the forage, and use as a turf crop. Some problems that have received much recent attention are discussed below.

A. Fescue Toxicosis Syndromes

Tall fescue has many excellent agronomic traits and should be a high-quality forage based upon chemical composition usually associated with forage quality, such as nitrogen, cell walls, and lignin. However, the potentially high nutritive value to grazing animals is offset by various fescue toxicosis syndromes.[12] including

- **Fescue foot** — A gangrenous condition of the feet, ear tips, and tail.
- **Fat necrosis** — Hard masses of abdominal fat that cause digestive upset and calving problems.
- **Summer slump** — Poor weight gains, changes in appearance (shaggy-rough hair coat, excessive salivation, nervousness), intolerance to heat, reduced pregnancy rate, and lower milk production.

Poor grazing performance of cattle, sheep, and horses in summer has been associated with consumption of tall fescue infected with an endophytic fungus, *Acremonium coenophialum* Morgan-Jones and Gams (previously called *Epichloe typhina*). Animals that consumed the infected fescue had a lower feed dry matter intake and a reduced average daily body weight gain. A general rule of thumb for grazing steers has been an average decrease of about 0.1 lb gain each day for each 10% increase of the fungal infection of the fescue forage. Agalactia, or poor milk production by cows, ewes, and mares, has been recorded. Reduced prolactin levels also occur. Animal physiological symptoms include elevated body temperatures, excessive respiration rates, excessive salivation, reproduction problems, and poor blood circulation in feet and tails, as well as failure to shed winter coats, nervousness, and the desire to seek shade.

The poor animal performance from consumption of endophyte-infected tall fescue has been attributed to the production of toxin(s) per se, either by the endophyte stimulating tall fescue to produce the toxin or by synergistically producing the toxin with tall fescue.[19] Numerous alkaloid compounds have been shown to be produced by the endophyte in culture and/or isolated from endophyte-infected tall fescue. These compounds are reportedly either toxic, rare, or related to insect deterring mechanisms.[2]

The fungus is referred to as an endophyte because it lives entirely within the tall fescue host plant in a symptomless, nonsporulating, nonpathogenic, and compatible relationship.[19] Bacon and Siegel[2] report that the endophyte exists as a nonhaustorial intercellular mycelium. The mycelia are found within the intercellular spaces of leaf sheaths and flowering stems and not in leaf blades and roots. The fungal infection is distributed unequally in tall fescue, and during the vegetative periods in the spring and fall it is located in the meristem tissue of shoot apexes and in assimilative and storage tissue of leaf sheaths and crowns.[2] The fungus invades developing seeds and is transmitted through seed; thus, the recent emphasis on use of fungus-free seed to establish new pastures.

The endophyte-fescue relationship appears to be similar to mutualistic symbiotic relationships reported for certain mycorrhiza and lichens[2] and, thus, might be better expressed as an endosymbiont-grass relationship. The manipulation of such an endosymbiont holds promise for producing beneficial effects in tall fescue and other grasses infected with the fungus. That is, the presence of the fungus decreases the probability that the plant will be consumed by grazing animals, thereby reserving a host for the fungus. While the endophyte is undesirable in livestock forage, it may be advantageous in turfgrasses used for industrial parks, lawns, athletic fields, revegetation of roadsides and disturbed areas, and other recreational or conservation purposes.[20]

The outstanding value of fescue as forage and turfgrass appears to be a consequence of various physiological parameters associated with the plant. However, the extent to which the fungus can modify the basic physiological processes of the grass is not known.[2] Methodologies have been reported for infection studies of grasses with endophytes.[21-23] Such studies can provide information on the plant-fungus interaction and allow the characterization of beneficial and detrimental effects of infected plants.

Favorable agronomic features attributed to tall fescue include its wide geographic adaptation, ability to withstand grazing, tolerance to poor soil conditions, resistance to insects and diseases, and presistence during drought. To what extent these desirable characteristics are the consequence of a plant-fungus interaction remains to be clarified. Although laboratory and greenhouse evidence is cited for an insect- or nematode-deterring mechanism in endophyte-infected fescue plants, such a relationship remains to be verified under field conditions.[20,22,24] Tolerance to environmental stresses,[25] improved stand persistence,[25] fall recovery,[26] and resistance to weed invasion[26] have been attributed to endophyte-infected tall fescue. Others[24,27] have observed no decline in stand nor yield of endophyte-free tall fescue. Pedersen and Sleper[24] concluded that it is too early to predict what effects total elimination of the endophyte from tall fescue would have on stand persistence.

A goal for enhancing the beneficial properties of endophyte-infected tall fescue cultivars would be to modify the physiological effects to allow only the desired beneficial traits to be expressed. Bacon and Siegel[2] postulated that genetic engineering might allow the introduction of gene sequences for beneficial properties from either the host or the fungus.

Infestation levels of the endophyte fungus in tall fescue pastures are estimated to be between 60 and 90%.[20] Infestation level refers to the percentage of all plants within a field that contain the fungus. Since there are no outward visible signs of the fungus on fescue plants, laboratory tests are needed for rapid determination of the fungal infection.[28]

Replacement of endophyte-infected fescue in the animal diet results in increased gain of grazing yearling cattle, weaning weights of nursing calves, milk production, and conception rates of cows.[29] Although a threshold level below which adverse effects of fungus-infected plants are minimized has not been established, total replacement may not be necessary for pastures with less than 30% infected plants.[20] However, for dairy cows and breeding mares, fescue fields testing over 10% infected plants should be replaced. Reduced levels of a hormone, prolactin, have been identified in the blood of ruminants fed infected fescue.[12,30] Preliminary evidence has shown that a dopamine antagonist administered to a grazing steer will restore both its ability to gain weight and to maintain the levels of prolactin in its blood.[12,31]

A major emphasis at the present time is on management strategies for reducing level of infection through dilution by adding clover to the pasture mixtures, preventing seedset of the established sod, and by actual destruction of the infected sods and reestablishment using fungus-free seed.[20] Reducing the amount of toxins consumed by animals and increasing the net energy of the diet have been proposed for improving the performance of cattle ingesting infected fescue.[32] Stokes, et al.[32] found that cattle offered grain, along with infected fescue, consumed more digestible dry matter than those that received nontoxic fescue to supplement infected forage diets. Nevertheless, the major problem remaining with cool-season grasses is the lowered digestibility and reduced animal performance during the summer. Many factors associated with lower digestibility have not been resolved with the discovery of the fungal endophyte infecting tall fescue. Solution of these problems will involve genetic improvement of tall fescue, and use of cell and tissue culture offer promise for more rapid achievement of that goal.

B. Dry Matter Digestibility

Although tall fescue compares favorably in dry matter digestibility (DMD) to other cool-season forage grasses, its DMD declines from 70 to 80% during early spring and late fall to 60 to 65% during the summer.[8,33] Heritability of DMD is low,[34] possibly because several factors affect digestibility. These include environmental conditions, management, neutral detergent fiber (NDF), acid detergent fiber (ADF), and lignification.[13] A large number of genes control DMD, and the potential for rapid genetic improvement is low by traditional methods. In general, lignin content and lignification during the summer period is the greatest problem and thus may offer the greatest promise for improvement. There is a strong inverse relationship between lignin and DMD which makes lignin a valuable and reliable negative indicator of forage quality.[1] Preliminary studies suggest coumaric acid is a precursor of lignin. *In vitro* determinations of DMD have shown a negative correlation between digestibility and phenolic concentration.[35,36] The lignin molecule is a condensation product of phenolic monomers[37] and is bound to several carbohydrates through *para*-coumaric and ferulic acid esters. The actual effect of the phenolic content of forages on ruminant performance is not well documented. However, the negative correlation between *in vitro* DMD and phenolic concentration cannot be ignored. Studies need to be conducted to determine the number of genes controlling accumulation of phenolic compounds that affect dry matter digestibility of ruminants. Coumaric acid and/or lignin per se appear to be dependent on relatively few genes and, therefore, show high heritability, permitting some progress to be made in improving DMD.

Intergeneric and interspecific hybridization of *Lolium-Festuca* species and *Festuca* species, respectively, increased genetic variability of the population for ADF and digestibility.[1,38] This finding permitted the development of hybrid derivative cultivars "Kenhy" and "Johnstone" with improved digestibilities and animal performance.[30,39] The DMD of some breeding lines of the hybrids and their derivatives approach 68% during the summer.[40] Dry matter digestibility would not be a problem in cultivars that averaged 70% or higher during the summer. Variability of hybrid genotypes suggest that attaining levels of 70% digestibility is realistic.

There are a number of toxic factors affecting DMD, of which a fraction containing the alkaloid perloline has been found in certain lines of tall fescue. Although perloline was cited as being toxic to rumen microflora and to reduce animal performance,[41] the involvement of perloline as an antiquality component of forage crops needs further clarification. Perloline was found to be highest during the summer, when lignification is also highest.[8] Buckner, et al.[42] found perloline levels to be highly heritable and Cornelius, et al.[14] found the levels to be dependent upon a relatively few genes. These findings permitted development of the Johnstone cultivar, which is characterized by low levels of perloline.[42]

C. Mineral Metabolism

A general impairment of mineral metabolism in animals grazing tall fescue has been associated with low animal performance.[43,44] The low apparent availability of calcium (Ca), magnesium (Mg), phosphorus (P), and sulfur (S) observed in tall fescue relative to other cool-season grasses[43,44] may be in part related to mineral solubilization in the rumen.[45] However, van Eys and Reid[45] concluded that the influence of other processes associated with mineral availability, such as absorption, endogenous secretion, or nutrient interaction, must be considered as well.

Grass tetany, or hypomagnesemia, is a complex metabolic disorder of ruminants that continues to be a factor under grazing management systems where environmental conditions may trigger the syndrome.[46] Cool, wet weather in the spring is particularly condusive to tetany problems.

Potassium (K) levels are generally high in tall fescue, but some plants have low values.[40] Thus, genetic approaches may be effective in modifying K uptake of fescue plants. One approach suggests that K uptake is dependent upon recessive genes, and the low-K plants breed true, showing that the level of K can be altered through a breeding program.[40] A reduction in K concentration in fescue plants could modify the K/(Ca + Mg) ratio and thereby decrease the potential for the forage to induce grass tetany in grazing cattle. Use of tall fescue haploids, doubled haploids, and other cell and tissue culture approaches,[10,11,17,18] in conjunction with traditional plant evaluation, offers the opportunity for more rapid progress toward resolution of tall fescue problems.

V. SUMMARY

Tall fescue is well adapted to the U.S. transition zone and is the predominant cool-season pasture species of this vast region. The geographic location of the transition zone, concomitant with its climatic and edaphic factors, is ideal for both tall fescue and animal production.

Summer syndrome, a severe nutritional problem affecting ruminant animals, has been associated with tall fescue forage that is infected by the fungal endophyte *Acremonium ceonophialum*. Tall fescue cultivars with low perloline levels, and digestibilities of 70% or better during the summer period, should contribute measurably to making the U.S. transition zone one of the most intensive livestock-producing regions of the world.

Improvement of tall fescue by conventional breeding methodology is slow because it is a wind-pollinated, highly self-sterile perennial. Therefore, tissue and anther culture methods will be useful to more rapidly isolate and/or regenerate plants with specific genetic characteristics. Such procedures should allow elimination of undesirable and recombination of desirable characteristics in less time, followed by conventional field tests and evaluations to combine new and traditional approaches to more rapidly improve this important forage and turf grass.

REFERENCES

1. **Buckner, R. C. and Bush, L. P.,** *Tall Fescue,* American Society of Agronomy, Madison, WI, 1979, 1.
2. **Bacon, C. W. and Siegel, M. R.,** Endophyte parasitism of tall fescue, *J. Prod. Agric.,* 1, 45, 1988.
3. **Hanson, A. A., Juska, F. V., and Burton, G. W.,** Species and varieties, in *Turfgrass Science,* Hanson, A. A. and Juska, F. V., Eds., American Society of Agronomy, Madison, WI, 1969, 393.
4. **Burns, J. C. and Chamblee, D. S.,** Adaptation, in *Tall Fescue,* Buckner, R. C. and Bush, L. P., Eds., American Society of Agronomy, Madison, WI, 1979, chap. 2.
5. **Mays, D. A. and Washko, J. B.,** The feasibility of stockpiling legume-grass pasturage, *Agron. J.,* 52, 190, 1960.
6. Virginia Agricultural Experiment Station, Report of Forage-Livestock Field Day, Middleburg, VA, October 7, 1987, 38.
7. **Terrell, E.,** Taxonomy, morphology, and phylogeny, in *Tall Fescue,* Buckner, R. C. and Bush, L. P., Eds., American Society of Agronomy, Madison, WI, 1979, chap. 3.
8. **Asay, K. H., Frakes, R. V., and Buckner, R. C.,** Breeding and cultivars, in *Tall Fescue,* Buckner, R. C. and Bush, L. P., Eds., American Society of Agronomy, Madison, WI, 1979, chap. 7.
9. **Berg, C. C., Webster, G. T., and Jauhar, P. P.,** Cytogenetics and genetics, in *Tall Fescue,* Buckner, R. C. and Bush, L. P., Eds., American Society of Agronomy, Madison, WI, 1979, chap. 6.
10. **Kasperbauer, M. J. and Eizenga, G. C.,** Tall fescue doubled haploids via tissue culture and plant regeneration, *Crop Sci.,* 25, 1091, 1985.
11. **Kasperbauer, M. J., Karlen, D. L., and Burton, H. R.,** Ploidy effects on protein, in vitro dry matter disappearance, and Potassium/(Calcium + Magnesium) equivalent ratio in tall fescue forage, *Crop Sci.,* 27, 1081, 1987.
12. **Sanchez, D.,** The wonder grass: a mixed blessing?, *Agric. Res.,* September, 12, 1987.
13. **Buckner, R. C.,** The fescues, in *Forages The Science of Agriculture,* 4th ed., Heath, M. E., Metcalfe, D. S., and Barnes, R. F., Eds., Iowa State University Press, Ames, IA, 1985, chap. 25.
14. **Cornelius, P. L., Buckner, R. C., Bush, L. P., Burrus, P. B., II, and Byars, J.,** Inheritance of perloline content in annual ryegrass × tall fescue hybrids, *Crop Sci.,* 14, 896, 1974.
15. **Sleper, D. A., Garner, G. B., Asay, K. H., Boland, R., and Pickett, E. E.,** Breeding for Mg, Ca, K, and P content in tall fescue, *Crop Sci.,* 17, 433, 1977.
16. **O'Toole, J. C. and Bland, W. L.,** Genotypic variation in crop plant root systems, in *Advances in Agronomy,* Vol. 41, Brady, N. C., Ed., Academic Press, New York, 1987, 91.

17. **Kasperbauer, M. J., Buckner, R. C., and Bush, L. P.,** Tissue culture of annual ryegrass × tall fescue F_1 hybrids: callus establishment and plant regeneration, *Crop Sci.*, 19, 457, 1979.
18. **Kasperbauer, M. J., Buckner, R. C., and Springer, W. D.,** Haploid plants by anther-panicle culture of tall fescue, *Crop Sci.*, 20, 103, 1980.
19. **Bush, L. P. and Burrus, P. B., II,** Tall fescue forage quality and agronomic performance as affected by the endophyte, *J. Prod. Agric.*, 1, 55, 1988.
20. **Fribourg, H. A., Wilkinson, S. R., and Rhodes, G. N., Jr.,** Switching from fungus-infected to fungus-free tall fescue, *J. Prod. Agric.*, 1, 122, 1988.
21. **Funk, C. R., Halisky, P. M., Ahmad, S., and Hurley, R. H.,** How endophytes modify turfgrass performance and response to insect pests in turfgrass breeding and evaluation trials, in *Proc. 5th Int. Turfgrass Research Conf.*, Lemaire, F., Ed., Avignon, France, 1985, 137.
22. **Johnson, M. C., Dahlman, D. L., Siegel, M. R., Bush, L. P., Latch, G. C. M., Potter, D. A., and Varney, D. R.,** Insect feeding deterrents in endophyte-infected tall fescue, *Appl. Environ. Microbiol.*, 49, 568, 1985.
23. **Latch, G. C. M. and Christensen, M. J.,** Artificial infection of grasses with endophytes, *Ann. Appl. Biol.*, 107, 17, 1985.
24. **Pedersen, J. F. and Sleper, D. A.,** Considerations in breeding endophyte-free tall fescue forage cultivars, *J. Prod. Agric.*, 1, 127, 1988.
25. **Read, J. C. and Camp, B. J.,** The effect of the fungal endophyte *(Acremonium coenophialum)* in tall fescue in animal performance, toxicity, and stand maintenance, *Agron. J.*, 78, 848, 1986.
26. **Hurley, R. H., Funk, C. R., Halisky, P. M., Saha, D. C., and Johnson-Cicalese, J. M.,** The role of endophyte enhanced performance in grass breeding, in Proc. 28th Grass Breeder's Work Conf., College Station, TX, June 19 to 21, 1984.
27. **Siegel, M. R., Johnson, M. C., Varney, D. R., Nesmith, W. C., Buckner, R. C., Bush, L. P., Burrus, P. B., II, Jones, T. A., and Boling, J. A.,** A fungal endophyte in tall fescue: incidence and dissemination, *Phytopathology,* 74, 932 1984.
28. **Reddick, B. B.,** Detection of the tall fescue endophyte with emphasis on enzyme-linked immunosorbent assay, *J. Prod. Agric.*, 1, 133, 1988.
29. **Boling, J. A.,** Endophytic fungus and tall fescue utilization by ruminants, *Prof. Anim. Sci.*, 1, 19, 1985.
30. **Baxter, H. D., Owen, J. R., Buckner, R. C., Hemken, R. W., Siegel, M. R., Bush, L. P., and Montgomery, M. J.,** Comparison of low alkaloid tall fescues and orchardgrass for lactating Jersey cows, *J. Dairy Sci.*, 69, 1329, 1986.
31. **Lipham, L. B., Thompson, F. N., Stuedemann, J. A., and Sartin, J. L.,** Effects of metoclopramide on steers grazing endophyte-infected fescue, *J. Anim. Sci.*, 66 (Suppl. 1), 373, 1988.
32. **Stokes, S. R., Goetsch, A. L., Nejad, H. H., Murphy, G., Jones, A. L., Mashburn, S., Beers, K. W., Johnson, Z. B., and Piper, E. L.,** Effects of supplemental bermudagrass hay or corn on intake, digestion and performance of cattle consuming endophyte-infected fescue, *J. Anim. Sci.*, 66, 204, 1988.
33. **Buckner, R. C., Todd, J. R., Burrus, P. B., II, and Barnes, R. F.,** Chemical composition, palatability, and digestibility of ryegrass-tall fescue hybrids, 'Kenwell', and 'Kentucky 31' tall fescue varieties, *Agron. J.*, 59, 345, 1967.
34. **Buckner, R. C., Burrus, P. B., II, Cornelius, P. L., Bush, L. P., and Leggett, J. E.,** Genetic variability and heritability of certain forage quality and mineral constituents in *Lolium-Festuca* hybrid derivatives, *Crop Sci.*, 21, 419, 1981.
35. **Allinson, D. W. and Osbourn, D. F.,** The cellulose-lignin complex in forages and its relationship to forage nutritive value, *J. Agric. Sci.*, 74, 23, 1970.
36. **Burns, J. C. and Cope, W. A.,** Nutritive value of crownvetch forage as influenced by structural constituents and phenolic and tannin compounds, *Agron. J.*, 66, 195, 1974.

37. **Fahey, G. C., Jr., Al-Haydari, S. Y., Hinds, F. C., and Short, D. E.,** Phenolic compounds in roughages and their fate in the digestive system of sheep, *J. Anim. Sci.,* 50, 1165, 1980.
38. **Hemken, R. W., Bull, L. S., Boling, J. A., Kane, E., Bush, L. P., and Buckner, R. C.,** Summer fescue toxicosis in lactating dairy cows and sheep fed experimental strains of ryegrass-tall fescue hybrids, *J. Anim. Sci.,* 49, 641, 1979.
39. **Buckner, R. C., Boling, J. A., Burrus, P. B., II, Bush, L. P., and Hemken, R. W.,** Registration of Johnstone tall fescue, *Crop Sci.,* 23, 399, 1983.
40. **Buckner, R. C.,** Personal communication, 1986.
41. **Boling, J. A., Bush, L. P., Buckner, R. C., Pendlum, L. C., Burrus, P. B., Yates, S. G., Rogovin, S. P., and Tookey, H. L.,** Nutrient digestibility and metabolism in lambs fed added perloline, *J. Anim. Sci.,* 40, 972, 1975.
42. **Buckner, R. C., Bush, L. P., and Burrus, P. B., II,** Variability and heritability of perloline in *Festuca* sp., *Lolium* sp., and *Lolium-Festuca* hybrids, *Crop Sci.,* 13, 666, 1973.
43. **Powell, K., Reid, R. L., and Balasko, J. A.,** Performance of lambs on perennial ryegrass, smooth bromegrass, orchardgrass and tall fescue. II. Mineral utilization, in vitro digestibility and chemical compostion of herbage, *J. Anim. Sci.,* 46, 1503, 1978.
44. **Spears, J. W., Ely, D. G., and Bush, L. P.,** Influence of supplemental sulfur on in vitro and in vivo microbial fermentation of Kentucky 31 tall fescue, *J. Anim. Sci.,* 47, 552, 1978.
45. **van Eys, J. E. and Reid, R. L.,** Ruminal solubility of nitrogen and minerals from fescue and fescue-red clover herbage, *J. Anim. Sci.,* 65, 1101, 1987.
46. **Grunes, D. L. and Rendig, V. V.,** *Grass Tetany* (ASA Spec. Publ. 35), American Society of Agronomy, Crop Science Society of America, and Soil Science Society of America, Madison, WI, 1979, 1.

Chapter 2

POTENTIAL BIOTECHNOLOGICAL APPROACHES

Karan Kaul

TABLE OF CONTENTS

I. INTRODUCTION

Conventional plant breeding and selection approaches have been very successful for the genetic improvement of many species of crop plants. In general, the approach involves identification of plants that are superior for characteristics such as disease resistance, adaptation to a geographic region, and composition of seed and/or leaves, all of which contribute to the quality and quantity of the plant product. Basically, the process involves identification of desirable genes and incorporating them into plants that already have other desired characteristics. This involves identification of appropriate starting materials (often from plant introductions or other sources of genetically diverse germplasm), controlled cross-fertilization, appropriate back crosses, and much selection among progeny of the crosses. With self-fertile species, a number of selfed generations can result in highly inbred lines which can be crossed to produce superior hybrids, or they can be used in development of new cultivars.

Although the conventional approach to plant breeding has been very successful, the procedures require considerable time and land for even the self-compatible and easily cross-compatible species. Other species, such as tall fescue (*Festuca arundinacea*), are wind pollinated and highly self-infertile. This combination results in a great challenge to the plant breeder because the plants are usually heterozygous, and self-infertility rules out inbreeding to concentrate desired genes. While naturally occurring genetic differences among gametes from the same tall fescue plant can be an obstacle in conventional plant breeding, it could provide an important advantage for some of the biotechnological approaches that depend on genetic diversity among cells prior to regeneration of plants from such cells. Thus, it is possible that some of the biotechnological approaches might be useful to speed up identification of superior genes and to develop pure lines that can be used directly, or, in lieu of inbred lines, in development of hybrids or new synthetic cultivars of tall fescue.

The purpose of this chapter is to highlight various biotechnological approaches which are potentially useful for improvement of tall fescue. The term biotechnology is used in a broad sense and includes various aspects of plant cell and tissue culture that might speed up development of a superior genotype.

II. HISTORICAL BACKGROUND

The concept of *in vitro* propagation of plant tissues was put forth by Haberlandt in the beginning of the 20th century.[1] However, his pioneering attempts met with little success. Unavailability of an appropriate nutrient medium and his choice of fully differentiated mesophyll tissue as explants both contributed to failure of Haberlandt's early attempts to culture plant tissues *in vitro*. The earliest successful growth of plant tissues *in vitro* was achieved by Robbins and Kotte in the 1920s.[2,3] Both grew excised root tips in nutrient media. White and Gautheret, the founders of the modern discipline of plant tissue culture, worked with excised tomato root tips and carrot root tissue, respectively.[4,5] They demonstrated that excised roots could be grown indefinitely under *in vitro* conditions with appropriate aseptic nutrient media and timely subculturing.[4,5]

Early studies by Steward showed that quiescent cells in potato tubers could be stimulated to metabolize, grow, and divide under appropriate *in vitro* conditions.[6] The quest for appropriate tissue in which to study *in vitro* rejuvenation of nonembryonic tissue led to the use of carrot (*Daucus carota*) root phloem parenchyma and tobacco (*Nicotiana tabacum*) stem pith tissue as the two model systems for study of plant cell and tissue growth *in vitro*.[6,7]

Studies on carrot root tissue demonstrated that chemical ingredients present in liquid endosperm of coconut induced both unorganized and organized growth *in vitro*.[8] The organized growth resulted in formation of somatic embryos which had the growth competence of zygotic embryos and could give rise to whole normal plants, thus demonstrating the totipotentiality of the somatic carrot cells under appropriate conditions.[9]

Studies with the tobacco system by Skoog, Miller, and colleagues led to the discovery of kinetin and the role of auxin/cytokinin balance in the regulation of organogenesis from undifferentiated tissue.[10,11] These studies were extended by Murashige and Skoog to the formulation of a chemically defined tissue culture medium[12] which, with small modifications, has served as the basic tissue culture medium for a wide range of plant tissues.

With development of a chemically defined medium, the basic framework for plant tissue culture was complete and *in vitro* tissue culture could be used as a tool to study biological processes and to develop a rapid means of plant propagation and modification. As time passed, new taxa of plants could be grown in tissue culture and techniques such as embryo culture, single cell culture, protoplast culture, somatic hybridization, and genetic engineering were developed. Each phase in the evolution of cell and tissue culture methodology has been accompanied by widespread interest and extension of concepts to many plant species and potential uses. Often the research was done to obtain basic information. However, some of the recent procedures have been done in conjunction with plant breeding programs.

III. POTENTIAL APPROACHES

Monocotyledonous plants in general and grasses in particular were very difficult to grow and propagate in tissue culture until about 2 decades ago. Modified media formulations and better explants, however, resulted in procedures for the establishment of cell and tissue cultures of several monocot taxa and their manipulation to regenerate plants *in vitro*.[13-17]

A. Somatic Cell and Tissue Culture

Somatic cell and tissue cultures may be started using plant parts such as embryos, cotyledons, hypocotyls, roots, stems, leaves, and inflorescences as the explants. The exact choice of explant tissue varies from species to species and is often based on the objective of the research. The selected explant is usually surface sterilized by treating it with a disinfectant such as 70% ethanol, dilute sodium hypochlorite, dilute mercuric chloride, or hydrogen peroxide followed by thoroughly rinsing the explant with sterilized distilled water to remove residual disinfectant. After surface decontamination, the explants are cut (if necessary) into appropriate-sized pieces under aseptic conditions (usually in a sterile transfer hood). The explants are placed on nutrient media and incubated under controlled environments. Various nutrient media formulations have been used for successful establishment of tissue cultures.[18] One usually arrives at a suitable combination of explant and culture medium based on scientific literature and personal experience. Under suitable conditions of organic and inorganic nutrition, growth regulators, light, and temperature, certain cells of the explant proliferate to give rise to an undifferentiated, parenchymatous mass called callus tissue. Callus tissue can be maintained by periodic transfer to fresh nutrient medium and can accomplish one or more of the following objectives, usually by manipulating the auxin/cytokinin ratio in the medium.

1. Mass Clonal Propagation

The objective is to cause somatic cells to increase in number and to then regenerate into numerous identical plants. If completely successful, all regenerated plants will be genetically identical with each other and with the original explant source plant. The procedure can allow relatively rapid clonal replication with a small investment of time and space relative to that used when clonal propagation is via tillers from whole plants. From a practical standpoint, however, the growth stage of the explant tissue, the number of subcultures after callus establishment, and the growth regulator balance in the medium can all contribute to nonuniformity among the cells in culture and to the plants regenerated from those cells. Thus, plants regenerated from somatic tissue may be genetically uniform or nonuniform, depending on tissue source and culture conditions. Genetic variants derived from the same original somatic explant have been called somaclonal variants. The tissue culture-derived plants can be grown to maturity and evaluated under appropriate greenhouse and/or field conditions. Clonal propagation *in vitro* can occur by organogenesis and somatic embryogenesis.

a. Organogenesis

Single cells or a group of cells in callus tissue can differentiate to form roots or shoots, depending on the auxin/cytokinin ratio in the culture medium. When organogenesis results in shoot formation, such shoots can be excised from the callus and induced to form roots, resulting in rooted plantlets which can be transplanted to soil and grown outside the culture vessel. Plantlet formation via organogenesis has been reported in many taxa including several members of the family Poaceae.[19,20] Lowe and Conger initiated callus on explants from tall fescue and then regenerated shoots from the callus.[21] Other investigators established callus cultures from meristematic tissue from a F_1 hybrid of annual ryegrass × tall fescue.[22] They varied subculture number and duration before regeneration of shoots to influence the frequency of normal clonal plants and the frequency of somaclonal variants (see Chapter 4).

b. Somatic Embryogenesis

In this mode of differentiation, certain cells in the callus tissue start to behave like zygotes. These cells go through the ontogenetic stages of embryo development and eventually give rise to embryos which can germinate into plants. Somatic embryogenesis has been reported for a large number of taxa.[23] In recent years, somatic embryogenesis has been successful for clonal propagation of several grass species.[24] Embryogenic callus provides a potentially useful approach for mass clonal propagation of tall fescue. Production of genetically uniform tall fescue seeds by controlled breeding is extremely difficult because the species is self incompatible and wind-pollinated. Once techniques for *in vitro* embryogenesis from tall fescue tissue cultures are established, production of "artificial seeds" (somatic embryos enclosed in a gel capsule) should be possible. Theoretically, this approach could result in numerous "seeds" of a given genotype. About 20 different species of tall fescue are known to occur in the U.S. Interspecific hybrids can be obtained, but the hybrids are often sterile and cannot be propagated through seeds.[25] Thus, somatic embryogenesis and artificial seeds may prove useful for mass propagation of such hybrids. Recent progress in somatic embryogenesis in several grass species is discussed in Chapter 3.

2. Selection at the Cellular Level

Both spontaneous and induced mutations can occur among cells in tissue cultures. Mutant plants can then be regenerated from the relatively few genetically different cells in a large cell population being grown in culture. Some of the mutated cells are able to differentiate into variant/mutant plantlets because of the presence of tissue culture milieu. These variant/mutant plants can be established and used as a source of genetic diversity.

a. Mutations

Random mutations can be induced by exposing a large population of cells grown in culture to chemical mutagens or to radiation. Other mutations occur spontaneously during tissue culture. The resulting mutants can be screened against certain chemicals or disease-related toxins for desired characteristics.

b. Selection and Plant Regeneration

Plant tissue culture systems often result in a relatively large proportion of mutant or variant cells, usually after numerous subcultures. This may result due to genetic changes induced by the tissue culture medium or may be because cells with unusual genotype, which do not survive in nature, can thrive under the controlled tissue culture conditions. Whatever the reasons for this variability, plant tissue culture systems provide a useful avenue for isolation of biochemical mutants/variants. Threonine-rich mutants of corn have been obtained by tissue culture technique.[26,27] Perhaps tissue culture approaches may be of value in isolating nutritionally superior mutants of tall fescue.

Cell lines tolerant to certain kinds of stresses such as salinity, drought, and cold may be selected from a population of cells by exposing the population to increasing levels of the stress. These stress-tolerant cell lines may then be induced to form stress-tolerant plants. Stress-tolerant lines of several cultivated species, including some grasses, have been isolated by this technique.[28] Similar approaches might be applicable to tall fescue.

Disease-resistant plants can be regenerated from resistant cells. If a chemically defined toxin is associated with a certain disease, cell lines resistant to the toxin can be isolated by growing cell cultures in the presence of the toxin. Induction of plantlet formation from such cell lines can result in formation of disease-resistant plants. Corn plants resistant to southern corn leaf blight have been isolated from tissue cultures using the host-specific phytotoxin produced by the fungus *Helminthosporium maydis* as the selection agent.[29,30] Similar procedures might be used for isolation of disease-resistant tall fescue cells which can then be regenerated into plants.

B. Haploids and Doubled Haploids

Haploid cells are concentrated in anthers (which contain microspores and pollen) and ovules, which contain the egg cells. Haploid plants can be regenerated directly from immature pollen, or the proliferation of these cells can result in callus formation which can then be induced to differentiate into haploid plants. The first pollen-derived embryoids were reported in *Datura innoxia* by Guha and Maheshwari.[31,32] This was followed by culture of pollen-derived plants from a large number of taxa including several cereals and other grasses.[33-35] Kasperbauer et al. used a nurse tissue approach to obtain pollen-derived haploids from tall fescue.[36] The haploid plants were then cloned through tillers and evaluated under field conditions to identify superior haploids. Doubled haploid plants were obtained from selected haploids by induction of endomitotic divisions in tissue cultures derived from these haploid plants, followed by regeneration of plants from the doubled haploid cells.[37] This procedure enables one to obtain homozygous plants, which are fertile and should be useful in lieu of highly inbred lines in crosses with other doubled haploids to produce hybrids, or they might be used in multiline synthetics by crossing with other doubled haploids and/or with conventionally derived lines.

Both gynogenetic (derived from unfertilized egg) and androgenetic (derived from microspores or pollen) haploid tissue cultures have been obtained from several different taxa.[38] The haploid approach should be useful for obtaining doubled haploid tall fescue plants in which recessive genes are expressed. Also, fusion of protoplasts derived from haploid tissue cultures and/or plants might be useful for obtaining somatic hybrids with desired genotypes.

C. Protoplast Culture, Somatic Hybridization, and Genetic Engineering

The protoplast culture technique enables one to culture individual plant cells from which the cell wall has been removed (usually by controlled enzyme digestion). Such wall-less cells are optimally suited for cell fusion and genetic manipulation. Cell fusion between protoplasts can be attained in the presence of an appropriate medium. In this approach, it is essential that one can identify the fused cells that result from fusion of the desired protoplasts and that a plant can be regenerated from the fused protoplasts. Such somatic hybrids might be useful in interspecific and intergeneric hybridization involving tall fescue. Instead of the whole genome, it is also possible to introduce a small amount (such as one or a few selected genes) into protoplasts by techniques such as microinjection and electroporation to induce a process comparable to bacterial transformation. Genetically modified protoplasts resulting from somatic hybridization or transformation can be induced to form callus tissue which, in turn, can be induced to differentiate into plantlets. If totally successful, the regenerated plant would be genetically identical with the fused protoplast or the transformed cell from which it originated. Like many other areas of plant tissue culture research, the early experimental work in the area of protoplast culture was done with herbaceous dicots such as tobacco and petunia (*Petunia hybrida*).[39-41] Somatic hybridization and genetic engineering experiments have also been done primarily with dicotyledonous plants.[42,43] However, during the last decade, significant progress has been made in the area of monocot protoplast culture. Procedures for isolation and culture of protoplasts, callus formation, embryogenesis, and plant regeneration have been reported for some taxa of the family Poaceae.[44,45] Plant regeneration from protoplast cultures of tall fescue and two related species was reported recently.[46] As mentioned earlier, protoplast culture provides opportunities for genetic engineering at the cellular level. Genetic transformation of grass protoplasts has also been achieved.[47,48] Another potential use of protoplast fusion methodology is that of cybrid formation (fusion of a nucleated protoplast with an enucleated protoplast). This procedure may be suited for otherwise incompatible hybrids in which one hopes to keep the chromosomes of only one parent and the cytoplasm of both parents. As with the other biotechnological approaches discussed in this chapter, it is critical that the cybrid cell divides and regenerates into a whole plant.

D. Disease-Free Plants

Meristem tissue from diseased plants can be used to obtain disease-free plants from diseased ones.[49-51] The procedure is based on the observation that the youngest cells in the fast-growing meristematic regions of a plant are often free of viruses which are present in the older cells of the infected plant.[51] The virus-free, meristematic tissue is placed on plant regeneration medium, and, if successful, the regenerated plants are genetically identical to the source plant, but free of the virus. Virus-free plants of several taxa have been regenerated from meristem cultures.[49] Also, heat treatment of virus-infected plants followed by excision and culture of apical meristem has been used to obtain disease-free plants.[50] Such heat treatment followed by culture of excised meristem tissue might provide a useful procedure to rescue unique breeding lines of tall fescue that have become infected with the endophyte *Acremonium coenophialum*.

E. *In Vitro* Fertilization and Embryo Rescue

In vitro pollination, *in vitro* fertilization, and embryo culture have been used to overcome barriers to crossability and obtain hybrids which could not be obtained through conventional breeding procedures.[52-55] *In vitro* fertilization may be useful in crossing plants of two species or genera that are physically incompatible. However, the contributions of both parents would have to be sufficiently compatible to allow development of the embryo into a hybrid plant.

Embryo rescue is a useful approach following wide crosses that initiate embryo development but abort before the embryo completes development on the parent plant. The procedure allows the embryo to initiate on the host plant and to then be "rescued" (before it aborts) and grown into a plantlet on a suitable tissue culture medium.

IV. CONCLUDING REMARKS

Biotechnological approaches to tall fescue improvement can involve both induced or natural genetic variability at the cellular level, followed by regeneration of plants from the cells. Approaches to induce variability among cells include chemical mutagens, radiation, protoplast fusion, and microinjection of foreign genes into the cell. Natural variability among cells can be found among gametes from a heterozygous plant. Potentially, immature gametes can be used to form a haploid callus in which mutations and chemical screening can be accomplished. However, a potentially more direct approach is to culture haploid plants from the immature gametes, clone them (through tillers or somatic embryogenesis), and to evaluate the haploid plants under various field and controlled environments to identify superior haploids. By doubling the chromosome numbers in the better haploids, one should be able to obtain fertile, homozygous doubled haploids which are the genetic equivalent of highly inbred lines that can be crossed with other doubled haploids or with conventionally derived lines. While the natural heterozygosity of tall fescue is a distinct obstacle to conventional plant breeding approaches, it appears that the haploid/doubled haploid approach should be able to use this heterozygosity as a distinct advantage for more rapid improvement of this important forage and turf grass.

REFERENCES

1. **Haberlandt, G.,** Kulturversuche mit isolierten Pflanzenzellen, *Sitzungsber. Akad. Wiss. Wien.,* 111, 69, 1902.
2. **Robbins, W. J.,** Cultivation of excised root tips and stem tips under sterile conditions, *Bot. Gaz.,* 73, 376, 1922.
3. **Kotte, W.,** Kulturversuche mit isolierten Wurzelspitzen, *Beitr. Allg. Bot.,* 2, 413, 1922.
4. **White, P. R.,** Potentially unlimited growth of excised tomato root tips in a liquid medium, *Plant Physiol.,* 9, 585, 1934.
5. **Gautheret, R. J.,** Sur la possibilite de realiser la culture indefinie des tissus de tubercules de carotte, *C.R. Acad. Sci.,* 208, 118, 1939.
6. **Steward, F. C.,** Reflections on aseptic culture, in *Handbook of Plant Cell Culture,* Vol. 1, Evans, D. A., Sharp, W. R., Ammirato, P. V., and Yamada, Y., Eds., Macmillan, New York, 1983, 1.
7. **Skoog, F.,** Growth and organ formation in tobacco tissue cultures, *Am. J. Bot.,* 31, 19, 1944.
8. **Caplin, S. M. and Steward, F. C.,** Effect of coconut milk on the growth of explants from carrot root, *Science,* 108, 655, 1948.
9. **Steward, F. C., Blakely, L. M., Kent, A. E., and Mapes, M. O.,** Growth and organization in free cell cultures, *Brookhaven Symp. Biol.,* 16, 73, 1963.
10. **Miller, C. O., Skoog, F., Von Saltza, M. H., and Strong, F. M.,** Kinetin, a cell division factor from desoxyribonucleic acid, *J. Am. Chem. Soc.,* 77, 1392, 1955.
11. **Skoog, F. and Miller, C. O.,** Chemical regulation of growth and organ formation, *Symp. Soc. Exp. Biol.,* 11, 118, 1957.
12. **Murashige, T. and Skoog, F.,** A revised medium for rapid growth and bioassays with tobacco tissue cultures, *Physiol. Plant.,* 15, 473, 1962.
13. **Wilmar, C. and Hellendoorn, M.,** Growth and morphogenesis of *Asparagus* cells cultured in vitro, *Nature, (London),* 217, 369, 1968.
14. **Kaul, K. and Sabharwal, P. S.,** In vitro induction of vegetative buds on inflorescence segments of *Haworthia, Experientia,* 26, 433, 1970.
15. **Kaul, K. and Sabharwal, P. S.,** Morphogenetic studies on *Haworthia:* establishment of tissue culture and control of differentiation, *Am. J. Bot.,* 59, 377, 1972.
16. **Vasil, V. and Vasil, I. K.,** Somatic embryogenesis and plant regeneration from tissue cultures of *Pennisetum americanum* and *P. americanum* × *P. purpureum* hybrid, *Am. J. Bot.,* 68, 864, 1981.
17. **Ozias-Akins, P. and Vasil, I. K.,** In vitro regeneration and genetic manipulation of grasses, *Physiol. Plant.,* 73, 565, 1988.
18. **Gamborg, O. L.,** Plant cell cultures: nutrition and media, in *Cell Culture and Somatic Cell Genetics of Plants,* Vol. 1, Vasil, I. K., Ed., Academic Press, Orlando, FL, 1984, 18.

19. **Carter, O., Yamada, Y., and Takahashi, E.,** Tissue culture of oats, *Nautre (London),* 214, 1029, 1967.
20. **Ahloowalia, B. S.,** Forage grasses, in *Handbook of Plant Cell Culture,* Vol. 1, Ammirato, P. V., Evans, D. A., Sharp, W. R., and Yamada, Y., Eds., Macmillan, New York, 1984, 91.
21. **Lowe, K. W. and Conger, B. V.,** Root and shoot formation from callus cultures of tall fescue, *Crop Sci.,* 19, 397, 1979.
22. **Kasperbauer, M. J., Buckner, R. C., and Bush, L. P.,** Tissue culture of annual ryegrass × tall fescue F_1 hybrids: callus establishment and plant regeneration, *Crop Sci.,* 19, 457, 1979.
23. **Ammirato, P. V.,** Embryogenesis, in *Handbook of Plant Cell Culture,* Vol. 1, Evans, D. A., Sharp, W. R., Ammirato, P. V., and Yamada, Y., Eds., Macmillan, New York, 1983, 82.
24. **Vasil, I. K.,** Regeneration of plants from single cells of cereals and grasses, in *Genetic Engineering in Eukaryotes,* Lurguin, P. and Kleinhofs, A., Eds., Plenum Press, New York, 1983, 233.
25. **Barnes, R. F.,** Importance and problems of tall fescue, in *Biotechnology in Tall Fescue Improvement,* Kasperbauer, M. J., Ed., CRC Press, Boca Raton, FL, 1990, ch. 1.
26. **Hibberd, K. A., Walter, T., Green, C. E., and Gengenbach, B. G.,** Selection and characterization of a feedback-insensitive tissue culture of maize, *Planta,* 148, 183, 1980.
27. **Hibberd, K. A. and Green, C. E.,** Inheritance and expression of lysine plus threonine resistance selected in maize tissue culture, *Proc. Natl. Acad. Sci. U.S.A.,* 79, 559, 1982.
28. **Nabors, M. W.,** Increasing the salt and drought tolerance of crop plants, in *Current Topics in Plant Biochemistry and Physiology,* Randall, D. D., Blevins, D. G., Larson, R. L., and Rapp, R. J., Eds., University of Missouri, Columbia, 1983, 165.
29. **Gengenbach, B. G., Green, C. E., and Donovan, C. M.,** Inheritance of selected pathotoxin resistance in maize plants regenerated from cell cultures, *Proc. Natl. Acad. Sci. U.S.A.,* 74, 5113, 1977.
30. **Brettell, R. I. S., Goddard, B. V. D., and Ingram, D. S.,** Selections of Tms-cytoplasm maize tissue cultures resistant to *Drechslera maydis* t-toxin, *Maydica,* 24, 203, 1979.
31. **Guha, S. and Maheshwari, S. C.,** In vitro production of embryos from anthers of *Datura, Nature (London),* 204, 497, 1964.
32. **Guha, S. and Maheshwari, S. C.,** Cell division and differentiation of embryos in the pollen grains of *Datura* in vitro, *Nature (London),* 212, 97, 1966.
33. **Sunderland, N.,** Anther culture as a means of haploid induction, in *Haploids in Higher Plants Advances and Potentials,* Kasha, K. J., Ed., University of Guelph, Guelph, Ontario, 1974, 91.
34. **Maheshwari, S. C., Rashid, A., and Tyagi, A. K.,** Haploids from pollen grains — retrospect and prospect, *Am. J. Bot.,* 69, 865, 1982.
35. **Wenzel, G. and Foroughi-Wehr, B.,** Anther culture of cereals and grasses, in *Cell Culture and Somatic Cell Genetics of Plants,* Vol. 1, Vasil, I. K., Ed., Academic Press, Orlando, FL, 1984, 311.
36. **Kasperbauer, M. J., Buckner, R. C., and Springer, W. D.,** Haploid plants by anther-panicle culture of tall fescue, *Crop Sci.,* 20, 103, 1980.
37. **Kasperbauer, M. J. and Eizenga, G. C.,** Tall fescue doubled haploids via tissue culture and plant regeneration, *Crop Sci.,* 25, 1091, 1985.
38. **Lacadena, J.-R.,** Spontaneous and induced parthenogenesis and androgenesis, in *Haploids in Higher Plants Advances and Potential,* Kasha, K. J., Ed., University of Guelph, Guelph, Ontario, 1974, 13.
39. **Power, J. B. and Cocking, E. C.,** A simple method for isolation of very large number of leaf protoplasts by using mixtures of cellulase and pectinase, *Biochem. J.,* 111, 33, 1969.
40. **Otsuki, Y. and Takebe, I.,** Isolation of intact mesophyll cells and their protoplasts from higher plants, *Plant Cell Physiol.,* 10, 917, 1969.

41. **Usui, H. and Takebe, I.,** Division and growth of single mesophyll cells isolated enzymatically from tobacco leaves, *Dev. Growth Differ.,* 11, 143, 1969.
42. **Carlson, P., Smith, H. H., and Dearing, R. D.,** Parasexual interspecific plant hybridization, *Proc. Natl. Acad. Sci. U.S.A.,* 69, 2292, 1972.
43. **Power, J. B., Frearson, E. M., Hayward, C., George, D., Evans, P. K., Berry, S. F., and Cocking, E. C.,** Somatic hybridisation of *Petunia hybrida* and *P. parodii, Nature (London),* 263, 500, 1976.
44. **Lu, C., Vasil, V., and Vasil, I. K.,** Isolation and culture of protoplasts of *Panicum maximum,* Jacq. (Guinea grass): somatic embryogenesis and plantlet formation, *Z. Pflanzenphysiol.,* 104, 311, 1981.
45. **Vasil, V., Wang, D., and Vasil, I. K.,** Plant regeneration from protoplasts of *Pennisetum purpureum* Schum. (Napier grass), *Z. Pflanzenphysiol.,* 111, 233, 1983.
46. **Dalton, S. J.,** Plant regeneration from cell suspension protoplasts of *Festuca arundinacea* Schreb., *Lolium perenne* L., and *L. multiflorum* Lam., *Plant Cell Tissue Organ Cult.,* 12, 137, 1988.
47. **Lorz, H., Baker, B., and Schell, J.,** Gene Transfer to cereal cells mediated by protoplast transformations, *Mol. Gen. Genet.,* 199, 178, 1985.
48. **Potrykus, I., Saul, M. W., Petruska, J., Paszkowski, J., and Shillito, R. D.,** Direct gene transfer to cells of a graminaceous monocot, *Mol. Gen. Genet.,* 199, 183, 1985.
49. **Kartha, K. K.,** Elimination of viruses, in *Cell Culture and Somatic Cell Genetics of Plants,* Vol. 1, Vasil, I. K., Ed., Academic Press, Orlando, FL, 1984, 577.
50. **Kartha, K. K. and Gamborg, O. L.,** Elimination of cassava mosaic disease by meristem culture, *Phytopathology,* 65, 826, 1975.
51. **Hansen, J. A. and Hildebrandt, A. C.,** The distribution of tobacco mosaic virus in plant callus cultures, *Virology,* 28, 15, 1966.
52. **Blakeslee, A. F. and Satina, S.,** New hybrids from incompatible crosses in *Datura* through culture of excised embryos on malt media, *Science,* 99, 331, 1944.
53. **Raghavan, V.,** Embryo culture, *Int. Rev. Cytol.,* Suppl. 11B, 209, 1980.
54. **Kanta, K., Rangaswamy, N.S., and Maheshwari, P.,** Test-tube fertilization in flowering plants, *Nature (London),* 194, 1214, 1962.
55. **Rangaswamy, N. S. and Shivanna, K. R.,** Induction of gametic compatibility and seed formation in axenic cultures of diploid self-incompatible species of *Petunia, Nature (London),* 16, 937, 1967.

Chapter 3

SOMATIC CELL CULTURE AND EMBRYOGENESIS IN THE POACEAE*

D. J. Gray

TABLE OF CONTENTS

* Florida Agricultural Experiment Station Journal Series No. 9959.

I. INTRODUCTION

Our recently developed ability to select, nurture, and increase specific plant genotypes by somatic cell culture promises widespread application to crop improvement by providing means to expedite existing breeding efforts or by circumventing breeding programs altogether. New mass propagation methods may allow cloning of outstanding selections as an economical alternative to conventional seed and vegetative propagation. The value of somatic cell culture will vary from crop to crop due to a variety of factors including intrinsic crop value, method of crop production, existing level of cultivar development, and/or relative ease by which crop improvement can be implemented by conventional vs. somatic cell culture methods. However, nowhere is somatic cell culture of more immediate use than for crops such as tall fescue that are unimproved or difficult to improve due to genetic crossing barriers.

Tall fescue (*Festuca arundinacea* Schreb.) fits into this category because it is predominantly self-sterile, which makes development of parental inbreds or true-to-type seed impossible except for a few self-fertile lines. This creates a major roadblock to improvement because desirable traits from distinct populations cannot be combined into true-breeding lines. Due to low per-plant value, seed is currently the only means of stand establishment; therefore, outstanding selections cannot be vegetatively propagated at a commercially acceptable cost. Somatic cell culture offers a number of possibilities for circumventing these problems by providing means to select and increase desirable germplasm and to then clone outstanding individuals at realistic costs.

However, a problem of using somatic cell culture for tall fescue improvement is the paucity of existing methodology. The first report of regeneration from cultures of tall fescue was in 1979 when Lowe and Conger obtained plants from callus derived from mature embryos.[1] Plant regeneration occurred from 9.6% of new calli, but dropped to 3.1% by the third subculture. Haploid plants were subsequently obtained in 1980 from anther-panicle cultures.[2] In 1983, Dale and Dalton used immature inflorescences to regenerate plants both directly from the explant or indirectly from an intervening callus stage. They obtained an average of 10.2 plants per cultured explant.[3] Only recently, in 1988, regeneration of tall fescue plants from protoplast culture was reported.[4] Kasperbauer et al. regenerated plants from callus obtained from internodes and peduncles of an annual ryegrass (*Lolium multiflorum* Lam.) × tall fescue hybrid.[5] Each of these studies were similar in that a modification of Murashige and Skoog (MS) medium[6] with 2,4-D as a growth regulator was used. Dale and Dalton also showed that benzylaminopurine (BAP) enhanced plant regeneration.[3]

Although these reports demonstrated that plant recovery from somatic cell cultures of tall fescue and its hybrid was possible, in-depth studies centered on refining regeneration phenomena to a large-scale level have not yet been accomplished. For example, protocols to obtain plant regeneration from other culture types (including cell suspensions and protoplasts) exist in only one report from one laboratory.[4] Likewise, basic knowledge of regeneration modes possible for tall fescue (i.e., embryogenesis and/or organogenesis) is not available. Increasing our understanding of tall fescue somatic cell culture will be an absolute requirement before capabilities to improve germplasm via biotechnological procedures will be possible. Fortunately, a large body of literature dealing with somatic cell culture of other grasses and cereals already exists, and many similarities between tissue culture performance of various poaceous species, including tall fescue, can be drawn. The purpose of this chapter is to discuss methods of somatic cell culture and selection already in place for other grasses and cereals that could be of use for tall fescue improvement.

II. MODES OF REGENERATION

Several distinct types of regeneration have been reported from grass and cereal tissue cultures, including production of axillary buds from organized explant tissue, *de novo* shoot organogenesis from callus cultures, and, in the preponderance of examples, somatic embryogenesis from both organized and unorganized tissues. Although somatic embryogenesis is now well documented in members of the Poaceae, the existence of organogenesis is not universally accepted. This is due in part to the difficulty of clearly documenting organogenic regeneration by both ruling out the existence of rudimentary or abnormal forms of embryogenesis where recognizable somatic embryos are not formed and in defining significant developmental differences between organogenesis and incomplete embryogenesis where only shoot meristems are found. Thus, it is a distinct possibility that some reports of shoot organogenesis from callus cultures fail to recognize somatic embryogenesis, particularly where plants grow from poorly developed, precociously germinated embryos. It is also possible that certain forms of organogenesis and embryogenesis do not significantly differ from each other.

The confusion between embryogenesis and organogenesis is made more apparent by considering existing literature. Studies concerning a number of poaceous species reported either organogenesis or left the question of regeneration mode up to speculation.[3,7-23] However, many of these tended to be chronologically early investigations conducted at a time when embryogenesis had not yet been documented or accepted. Thus, relatively small, nondescript monocotyledonous somatic embryos may have been overlooked. This view is supported by literature for several species, including corn (*Zea mays* L.) where an early first report suggested organogenesis,[13] whereas later reports of the same system by the same laboratory confirmed embryogenesis.[24] Photographs of callus presented in many of the above cited papers strongly resemble embryogenic type callus (see Section III.B.1 for callus description). Additionally, nearly all studies of grass and cereal tissue culture employed a strong phenoxy herbicide such as 2,4-D to induce regeneration. Interestingly, these herbicides are the most commonly used growth regulators to obtain embryogenesis in plants in general. It is unlikely that use of the same growth regulators resulted in entirely different regeneration pathways. One of few exceptions to use of phenoxy herbicides was with rice (*Oryza sativa* L.) panicle culture, where NAA and kinetin induced plant regeneration; the authors postulated that adventitious buds occurred by reversion of floral primordia to a vegetative stage.[16] Reports of organogenesis occurring in the same cultures and in concert with embryogenesis[25-27] are especially difficult to confirm since forms of both normal and abnormal embryogenesis may coexist and lead to misinterpretation. Reports of organogenesis where coleoptilar development occurs[26] are clearly misinterpreted since coleoptiles are of embryonic and not shoot apical origin (see Section III.C on embryo ontogeny, below).

Finally, in considering modes of regeneration, it is important to focus on actual developmental differences between organogenesis and embryogenesis. Until relatively recently, the traditionally held conceptual difference between organogenic and embryogenic regeneration was that shoots were thought to develop from groups of cells with consequential adventitious rooting, whereas embryos arose from single cells and gave rise directly to bipolarly complete plants. For practical purposes, these differences suggested important consequences because organogenic regeneration would tend to require more culture steps to accomplish complete plant development and the plants would be more prone to chimeric abnormalities due to multicellular origins. In contrast, embryogenesis would require less culture manipulation to achieve plant regeneration, and chimeras resulting from pre-embryo cell division would not be possible. Additionally, development of useful nonchimeral mutants would be possible by manipulation and/or selection of the embryogenic cells. However, these concepts underlying the cell origin distinction between organogenic and embryogenic regeneration modes have more recently broken down. For example, it is now suggested that somatic embryos may develop both from single cells and by budding from multicelled masses (proembryonal complexes),[28] and, conversely, adventitious buds may arise from single cells.[29] More evidence for single-cell origins of organogenic shoots comes from transformation studies of tobacco (*Nicotiana tabacum* L.), where nonchimeral transformants were recovered from a classical organogenic culture system.[30] Therefore, although plant recovery via embryogenesis is more efficient than organogenesis, other reasons for favoring one over the other are less clear. Subsequent discussions in this chapter concerning plant regeneration in the Poaceae will deal primarily with somatic embryogenesis, since its occurrence is well documented.

III. SOMATIC EMBRYOGENESIS

A. Methodologies

Although the first successes in obtaining embryogenesis in the Poaceae required specific, strictly adhered-to explants and medium formulations, historically, the technological advancement of this field as evidenced by the literature base has shown that many distinct methodologies are useful. This section discusses explants and media proven successful for producing embryogenic cultures from various grasses and cereals.

1. Explants

Immature zygotic embryos of *Sorghum bicolor* (L.) Moench. were the first (1977) clearly successful explants used to initiate embryogenesis from a poaceous species.[31] Interestingly, a much earlier study (1962) of cultured immature barley (*Hordeum vulgare* L.) embryos described an "outgrowth of scutellum of a laminar nature" that, in retrospect, appears to have been a type of embryogenic callus.[32] However, the use of immature embryos for *S. bicolor* may have been stimulated by a more contemporary study of corn[13] that did not recognize regeneration via embryogenesis. Immature embryos remain the most commonly used explant for initiation of embryogenic cultures, with reports of success for over ten species.[33-41] Immature embryos presumably contain either an abundance of embryogenic cells or cells that can be induced in culture to become embryogenic. The embryos are obtained at various times after pollination or, alternatively, at various size parameters. For example, rye (*Secale cereale* L.) embryos were harvested 14 to 21 d after anthesis.[34] Mature embryos or intact caryopses have also been used to initiate regenerative cultures.[25,35,42-48] Generally, immature or mature embryos are excised from seeds and placed on solidified medium such that the shoot-radicle axis faces down and the scutellum up. This configuration encourages callus production from the scutellum.[13,31,38]

A serious limitation to practical use of embryo explants is their zygotic origin; resulting cultures are sexual progeny, not clones of outstanding, tested individuals. Use of explants from mature, performance-tested material is desired, and this has lead to exploration of other potential explant sources. Successful generation of embryogenic cultures from leaves of *S. bicolor* in 1980[52] represented a significant breakthrough in our ability to clone outstanding specimens. This advancement subsequently stimulated research to demonstrate that young leaves of other species were capable of embryogenesis.[49-59] Figure 1A shows callus proliferation from a cultured orchardgrass leaf.

Because grass leaves grow from their base, a large source of meristematic tissue with embryogenic potential exists near the attachment of leaf to stem. The leaf meristems are more pronounced and active in younger leaves, and relatively more meristematic cells exist in leaf tissue located proximal to the

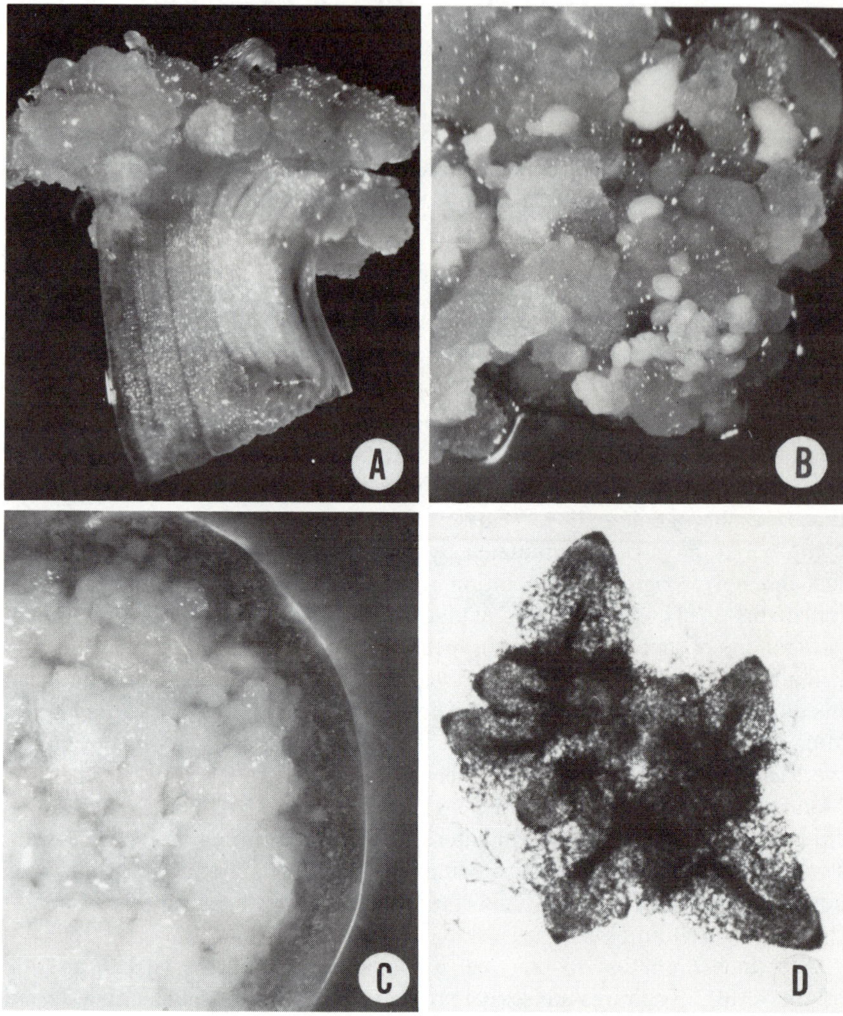

FIGURE 1. Orchardgrass callus types. (A) Initiation of heterogenous callus from a cultured leaf section; (B) friable embryogenic callus. Note white somatic embryos; (C) wet-type callus; (D) proliferating roots dissected from callus. ([B and C] from Gray, D.J., Conger, B. V., and Hanning, G. E., *Protoplasma*, 122, 196, 1984. With permission.)

stem. Thus, it has been noted that the more basal regions of younger leaves are most responsive.[52-55,60,61] Although leaves represent a true clonal tissue source, a limitation to their use remains for annual cereals because only basal meristems from very young plants will respond.[49,53,54,58,59,61] In comparison, young leaves of mature perennial grasses are obtainable from stock plants over long time periods.[50,51,55,56] This difference in response is undoubtedly due to the determinant growth pattern of cereal leaves vs. the indeterminant growth of grass leaves. Therefore, it is still not possible to use leaf culture for cloning of field-tested cereal genotypes.

Besides zygotic embryos and leaves, the third most commonly used explants for poaceous embryogenesis are various floral organs. Embryogenesis from 1- to 50-mm-long immature inflorescences of ryegrass[62] constituted the first successful example; however, organogenesis was also suggested in cultures that had no obvious embryo-like structures (see Section II for discussion of regeneration modes). Subsequently embryogenesis was obtained from immature inflorescences and/or panicles of several grass species,[63-68] in addition to the cereals *Setaria italica* (L.) Beauv.[69] and an interspecific *Oryza* hybrid.[70] Little is known of the embryogenic cell origin from such relatively complex tissues. However, the recovery of haploid plants from anther-panicle culture of tall fescue,[2] as described in Chapter 5 of this book, demonstrates that regeneration from meiotic cells is possible. Because floral organs represent an abundant source of meristematic tissue that develops relatively late in the plant life cycle, these explants may provide means to clone field-tested individuals.

Embryogenesis has also been reported from other poaceous plant explants, including anthers (see Chapter 5 of this book), nodes,[62] pistils,[72] roots,[26] shoot tips,[71] and stems.[36] With, the exception of anthers, these represent preliminary reports based upon single species. Although such scattered successes do not have practical large-scale application at this time (1989), these studies demonstrate that cells from many different organs that occur throughout the grass plant life cycle can have embryogenic potential.

2. *Media*

It is not within the scope of this chapter to present in-depth analysis of medium components and their effects on somatic embryogenesis. However, to provide a well rounded view of basic methodologies for poaceous tissue culture, it is necessary to reference various common formulations, growth regulators, and special addenda that have figured prominently in successful studies.

The most commonly used nutrient formulation for plant tissue culture (including the Poaceae) in general is MS medium.[6] Schenk and Hildebrandt (SH) medium,[73] B5,[74] and N6[75] are also in common usage. For comparison of specific medium ingredients, see Reference 76. There have been relatively few direct comparisons concerning the effect of these media on embryogenesis. For wheat (*Triticum aestivum* L.), it was shown that SH medium supported less growth than MS medium.[77] Likewise, SH medium was not as good as Linsmaier and Skoog (LS) medium[78] for embryogenesis in corn[79] (LS medium is identical to MS medium except for vitamin constituents). Using corn, N6 was shown by one laboratory to be better than MS for production of friable embryogenic callus,[80] whereas another laboratory reported no difference.[81] However, N6 was clearly shown to be superior to MS for culture of corn protoplasts.[82] At present (1989), no general recommendation for basic medium formulation can be made; each should be tested on an individual case basis.

Many organic addenda to the above media have been tested and are too numerous[83] to be treated in depth here. However, several generalities can be drawn. Among organic nitrogen sources, L-proline resulted in increased friability and/or embryogenesis in maize.[24,80] Combinations of proline and serine initiated and promoted embryo development in suspension cultures of orchardgrass (*Dactylis glomerata* L.).[84] Growth of corn protoplasts was best in medium containing both proline and asparagine.[82] Tryptophan stimulated production of embryogenic cultures in rice.[85] Complex mixtures of organic nitrogenous compounds in the form of casein hydrolysate also stimulated embryogenesis.[24,81,86,87] Among vitamin constituents, a 5-times elevation of thiamine concentration was shown to stimulate percent initiation of friable embryogenic corn cultures.[81] Sucrose is typically incorporated at 30 g/l, although a number of studies have shown better culture growth and/or embryo maturation at 60 g/l.[24,80]

Growth regulator choices for initiation and maintenance of poaceous embryogenic cultures are relatively straightforward. In nearly all instances, a phenoxy herbicide with strong auxin-like activity is required. A great majority of studies employed 2,4-D;[79,81,88] however, dicamba,[55] 2,4,5-T,[48,85] and others have also been successfully used. Hanning and Conger demonstrated that 2,4-D was equivalent or superior to dicamba for orchardgrass embryogenesis.[60] However, dicamba was superior to 2,4-D for corn[83] and wheat.[77] The cytokinins, kinetin[48,77] and benzyladenine,[85] have more uncommonly been reported to stimulate embryo maturation and germination.

Although relatively high auxin concentrations (ca. 5 mg/l 2,4-D) are needed to initiate and maintain embryogenic cell lines, such concentrations often inhibit later stages of embryo development. Therefore, for embryo maturation and, subsequently, plant regeneration from established cultures, the auxin is typically either lowered (ca. 0.1 mg/l 2,4-D), replaced with a weaker auxin such as IAA or NAA, or omitted altogether. In some instances, however, embryogenesis and maturation occur at the same growth regulator concentration[55,87] so that only one medium is needed for initiation, maintenance, and embryo maturation. In all cases, high auxin concentration inhibits embryo germination so that levels must be decreased or omitted for plant regeneration.

B. Growth of Embryogenic Cultures

1. Recognition and Maintenance of Embryogenic Callus Cultures

During culture initiation and maintenance, grasses and cereals produce a number of different callus types. These types have been separated into two general categories, embryogenic and nonembryogenic, based upon whether

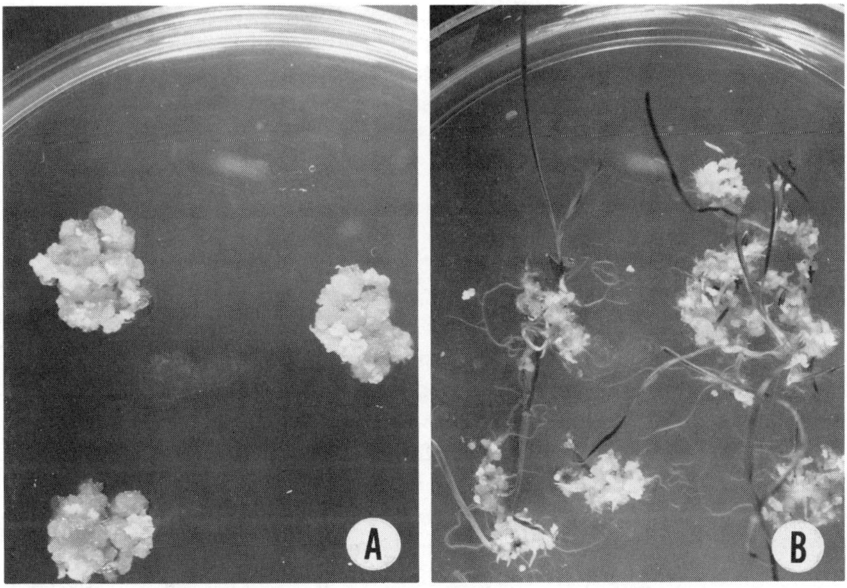

FIGURE 2. Effects of growth regulator on orchardgrass callus proliferation and embryo germination. (A) Friable embryogenic callus maintained on SH medium containing 30 μM dicamba. (B) Embryogenic callus on SH medium lacking growth regulators. Note numerous plantlets from germinated embryos.

or not somatic embryos are produced. Embryogenic callus either produces somatic embryos or is capable of embryogenesis when subjected to proper environmental conditions. Regardless of the presence or absence of preexisting somatic embryos, these two callus types can be separated on the basis of gross morphology. Embryogenic callus tends to be either: (1) compact, smooth, white, knobby,[46] and slow growing, being composed chiefly of fused, proliferating scutellar tissue,[35,38,80,89-92] or (2) highly friable with embryos interspersed among masses of dry, loosely organized cells (Figure 1B).[24,55,86] Nonembryogenic callus can also be separated into two distinct types: (1) translucent-to-yellow-to-brown (necrotic) callus with a wet (e.g., Figure 1C), rough, crystalline, unorganized appearance and composed of rather large elongated cells,[46] or (2) highly organized tissue composed of a loose callus matrix with proliferating root primordia (Figure 1D).[44,49,53,61,62,76,93,94]

Typically, embryogenic cultures are composed of both embryogenic and nonembryogenic sectors.[46] Often, the nonembryogenic callus proliferates at a greater rate than the embryogenic type; thus, it is possible to lose embryogenic sectors over time. Due to these unequal proliferation rates, it is essential to identify and manually isolate embryogenic callus during subculture. By repeated selection during transfers, cultures composed predominantly of embryogenic callus can be maintained (Figure 2A). Selection must be an ongoing process, however, since proliferation of the aforementioned cell and tissue types occur continually from embryogenic callus. For example, highly embryogenic orchardgrass callus becomes predominantly composed of proliferating root primordia after several transfers without selection (unpublished observation). Lowering or removing auxin during subculture results in embryo germination (Figure 2B) and also root formation.

2. *Suspension Cultures*

Suspension culture methodology for grasses and cereals has been slow in development due, in part, to the tendency of cultures to differentiate into a variety of undesirable cell types, as described above. Many early attempts at suspension culture initiation resulted in either growth of roots, nongrowing clumps, or proliferating undifferentiated cells with no morphogenetic capacity (see Reference 94 for discussion). However, refinement of protocols for production of highly embryogenic callus cultures has led to proliferate totipotent suspensions. Regenerative suspension cultures have now been reported for corn, guinea grass (*Panicum maximum* Jacq.), napier grass (*Pennisetum purpureum* Schum.), orchardgrass, pearl millet (*Pennisetum americanum* (L.) K. Schum.), rice, ryegrass, and sugarcane (*Saccharum officinarum* L.).[24,26,39,81,90,95-99] Important factors that led to success for each of these species were use of suitable inoculation material, specific steps for early initiation, and, of course, medium components.

Central in establishment of embryogenic suspension cultures is use of inoculum with regenerative potential. Most systems use embryogenic callus as inoculum.[24,39,81,90,95-98] Friable-type embryogenic callus[24,86,95] is best for initiation because it tends to disperse in liquid more readily, and this results in earlier establishment of the suspension culture. The compact, white, slow-growing embryogenic callus is much more difficult to establish. Typically, this type of callus is chopped or teased apart into smaller pieces prior to inoculation or placed *in toto* into the medium.[39,90,97,98] These relatively organized pieces of tissue then produce smaller proliferating cell masses as the liquid cultures are agitated. Inoculum density is critical in some instances. For corn, 2 g of friable callus was inoculated into 100 ml of medium, and cultures attained a stable growth rate after 2 weeks.[24] Rarely, suspension cultures have been established from organized tissues. Root tips of rice were inoculated into 125-ml flasks, resulting in regenerative cultures.[26] The organogenic regeneration mode described in that report may have been incorrectly interpreted (see Section II, above). Both zygotic[99] and somatic[86] embryos were also used to initiate embryogenic suspension cultures. Here, it was found that inoculum density as well as initiation procedures were important factors. Embryos were first cultured in small volumes (1.0 to 2.5 ml) of liquid medium. The medium volume was gradually increased as rapid scutellar proliferation occurred until eventually the resulting liquid suspension was inoculated into larger flasks on a rotary shaker and maintained through weekly transfer intervals.

Often, the initiation phase is relatively slow and requires special steps not needed in culture maintenance. Kamo and Hodges[95] found that corn suspension cultures could not be divided during the first month of culture; instead, the entire volume of medium was replaced each week. Thereafter, the cultures were diluted (1:1) weekly with fresh medium for an additional 2 months. Once established, the cultures were maintained and increased by diluting weekly in 5 volumes of fresh medium. Another protocol for initiation of corn suspension cultures involved establishment of cultures by pipetting the entire liquid volume from original inoculated flasks into new empty flasks.[81] This liquid contained newly produced and rapidly dividing cell clusters, but the original callus inoculum was left behind. The small clusters were allowed to settle prior to removal of the spent medium and replacement with fresh medium. This step was repeated weekly for 5 weeks and resulted in stable cultures.

Medium requirements are generally the same for suspension cultures as for callus culture except that a solidifying agent is not used. However, several addenda have been, variously, crucial to success. Auxin concentration must be critically adjusted. With corn,[95] for instance, low 2,4-D levels resulted in undesirable root proliferation, whereas high levels promoted culture browning. For orchardgrass, suboptimal dicamba levels not only promoted root growth, but result in less growth and less embryogenesis. High dicamba levels likewise inhibited growth and embryogenesis, but also resulted in increased culture viscosity, which was caused by a mucilaginous cellular secretion.[87] Abscisic acid, L-proline, and casein hydrolysate were helpful in corn suspension culture.[24,81,95] Casein hydrolysate exerted a profound effect on orchardgrass suspension cultures.[86,87] Orchardgrass suspension cultures grown without casein hydrolysate proliferated indefinitely, but did not undergo embryogenesis. However, embryogenesis was readily stimulated once high levels of casein hydrolysate (3 to 4 g/l) were added (Figure 3). Furthermore, with orchardgrass, the embryogenic response was reliably turned on and off indefinitely by addition or deletion of casein hydrolysate. Casein hydrolysate also interacts with auxin so that higher casein levels were needed to define optimum dicamba concentrations (Figure 3).[87] Recently, it was shown that a combination of proline and serine effectively substituted for the casein hydrolysate effect in orchardgrass.[84]

Growth rate of embryogenic suspension cultures is variable. Suspensions of orchardgrass doubled in mass approximately every 14 d,[86] whereas, cell dry weight of corn suspension cultures doubled every 48 h.[95] However, growth rate of established corn cultures was dependent on inoculum density. Cultures inoculated with 0.07 g dry weight of established suspension material reached early log phase growth and entered a stationary growth phase 1 to 2 d earlier than with lower inoculum densities. Furthermore, subculture using inoculum from mid log phase cultures resulted in best growth.[95] Embryogenic grass and cereal suspension cultures increase by the growth and fragmentation of

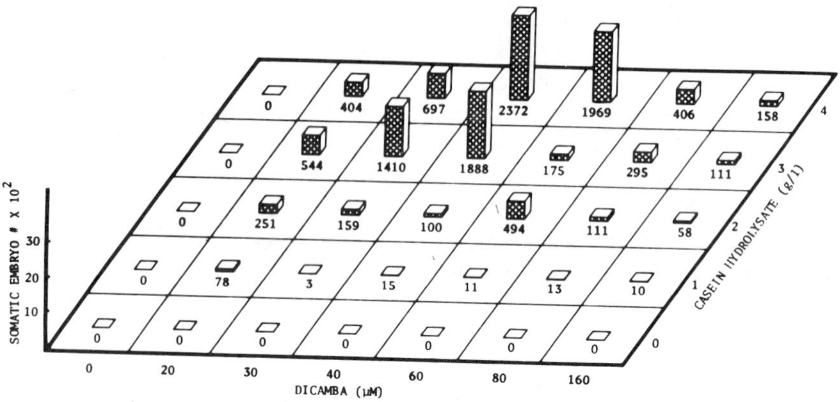

FIGURE 3. Effects of dicamba and casein hydrolysate on embryogenesis in orchardgrass suspension cultures. Data represents an average number of embryos per 35 ml of liquid collected 6 weeks after inoculation. (From Gray, D. J. and Conger, B. V., *Plant Cell Tissue Organ Cult.*, 4, 123, 1985. Reproduced by permission of MTP Press Limited, Lancaster, England.)

cell masses.[76,86,95] However, there is no evidence to suggest that single cells sloughed from masses are capable of sustained division and culture increase. Thus, it is critical to selectively transfer cell masses during subculture. In some instances, novel transfer procedures are required in order to maintain an adequate concentration of cell masses. Although the traditional method of pipetting inoculum from established cultures into fresh medium has been successful,[81] often this procedure does not allow transfer of suitable-sized cell masses due to the restricted pipette orifice. More commonly, an aliquot of established suspension material is diluted with fresh medium. For example, Kamo and Hodges[95] diluted one part of corn suspension culture with five parts fresh medium. For rice, the dilution ratio was either 1:2[26] or 1:4.[100] Transfer by dilution of relatively large aliquots allows cell masses to be maintained at the high concentrations needed for rapid culture increase.

Despite the notable progress in initiating and maintaining poaceous embryogenic suspension cultures, a remaining problem is that, with the exception of orchardgrass,[86,87] actual development of morphologically identifiable somatic embryos does not occur in liquid medium.[24,39,81,90,95,96,98] In these systems, embryos develop only after suspension culture-derived material is plated onto solidified medium. Thus, the presence of an embryogenic suspension culture is confirmed only after subsequent growth as a callus culture. Furthermore, plants are obtained only after several tedious manipulations. This limits the present usefulness of such suspension cultures for high-yield production of somatic embryos. Only in suspension cultures of orchardgrass can somatic embryos capable of germination be reared in a single liquid medium formulation (Figure 4A).[86,87] Orchardgrass embryogenesis is stimulated in suspension culture by the presence of casein hydrolysate[86] or a combination

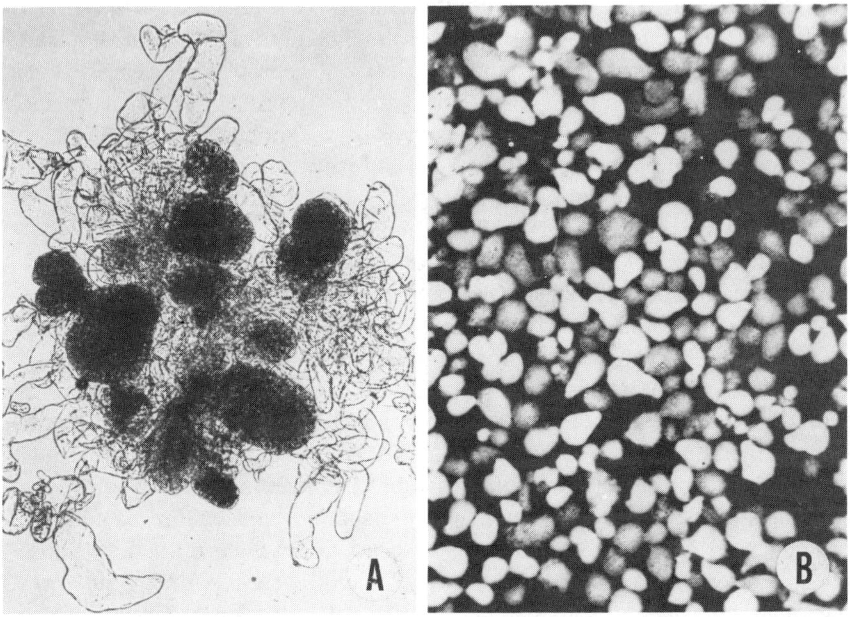

FIGURE 4. Embryogenic orchardgrass suspension culture. (A) Embryos differentiating in a suspended cell mass. (B) Sample of embryos collected from a suspension culture by seiving. (From Gray, D. J., Conger, B. V., and Hanning, G. E., *Protoplasma*, 122, 196, 1984. With permission.)

of proline and serine,[84] as described above. Therefore, the effects of various media and cultural manipulations on the complete process of embryogenesis in suspension cultures of the Poaceae can presently be studied only with this species. In orchardgrass, optimization of dicamba and casein hydrolysate levels to 40 μM and 4 g/l, respectively, resulted in the production of over 2300 recognizable embryos in 6 weeks from 1 ml of inoculum (Figure 3).[87] Well-developed embryos were collected *en mass* on a 0.18-mm mesh screen (Figure 4B).[86] Such embryos were induced to germinate by placement either in a thin layer of liquid or on solid medium lacking dicamba,[86] as well as by placement in nonsterile water or sand (unpublished observation). The high yield and ease of isolation of suspension culture-derived orchardgrass embryos vividly demonstrates the potential of using such systems for mass *in vitro* cloning, as described in Section IV.C. Development of equivalent systems for other poaceous species would be highly desirable. Advancement of cloning technology for these species will depend strongly on our ability to refine suspension cultures.

3. Protoplast Cultures

Regeneration from plant protoplasts is undoubtedly the most difficult means of obtaining plants by somatic cell culture. This is due to several factors including the difficulty associated with obtaining adequate yields of regenerable cells suitable for protoplast culture, the inherent fragility of protoplasts, and the sensitive nutritional requirements for wall regrowth and sustained cell division. Regeneration of mature plants from protoplasts of poaceous species has met with little success, despite significant emphasis by a number of laboratories. However, recent progress has resulted in plant regeneration for several cereals. Furthermore, a body of literature documenting generally applicable advances in technology has accumulated.

The first report of embryogenesis from poaceous protoplasts was in 1973 when formation of proembryos and, apparently, plants from suspension culture-derived protoplasts of bromegrass (*Bromus inermis* Leyss) was documented.[101] This report was chronologically early when compared to other literature on regeneration phenomena. The procedure was apparently not repeated in intervening years, and the paper has been largely ignored in subsequent literature. However, this early use of embryogenic suspension culture-derived cells for protoplast source material has been adopted by most subsequent studies. In 1979, procedures for isolation and culture to callus of protoplasts from suspension cultures of pearl millet were refined.[102] Subsequently, embryogenesis and plantlet regeneration were reported.[41] However, the establishment of plants capable of survival in soil was not successful. Since these early reports, similar results of embryogenesis and germination to plantlets, but not establishment of mature plants in soil from suspension culture[41,97,103,104] and callus culture-derived[105] protoplasts, have become relatively commonplace. The similarity of these results suggest a serious barrier to regeneration of grasses and cereals from single cells.

Only recently has clear documentation of plant regeneration from protoplasts become available for rice,[100,106,107] sugarcane,[108] and corn.[82] Most recently, plants were obtained from protoplasts of tall fescue and both annual and perennial (*Lolium perenne* L.) ryegrass.[4] Each of these reports employed embryogenic suspension cultures as source material except for Reference 106, which also utilized callus. However, each differed as to enzyme mixtures used and specific cultural details. All enzyme mixtures contained cellulase RS, but for rice, one report supplemented only one additional enzyme (pectolyase Y-23),[100] whereas another used both of these in addition to macerozyme R-10 and driselase.[106] For sugarcane, cellulase RS was supplemented with crude pectinase, driselase, and macerozyme.[108] Enzymes used for corn protoplast isolation were cellulase RS, pectolyase Y-23, and driselase.[82] These protocols show clearly that both a cellulase and pectinase are needed; however, it appears that several different preparations are suitable. A potential problem encountered with these relatively impure commercial enzyme preparations is the presence of salts, proteases, and other contaminants that may be harmful to fragile protoplasts. In some instances, enzymes are desalted prior to use,[82] resulting in enhanced cellulolytic activity and protoplast viability. Contaminating proteases may be inactivated by addition of bovine serum albumin[82] or calf serum.[106]

Protoplast culture methods also varied among these reports. For corn, protoplasts were cultured on filter papers over rapidly growing "feeder" cells.[82] The feeder cells apparently provided specific nutrient factors such that a relatively unmodified basal medium was sufficient for protoplast division (see Section III.A.2). For rice and sugarcane, protoplasts were cultured on complex media containing a number of organic constituents so that feeder cells were not necessary.[100,106,108] Immobilization of protoplasts in drops of agarose enhanced division frequency and allowed the surrounding liquid culture medium to be easily replenished and modified.[97]

Despite these recent breakthroughs in regenerating plants from protoplasts, a number of serious problems remain. Plant recovery and viability are still very low in most of these reports. For example, only 23 corn plants from 1 line and 2 from another matured in soil, and all were sterile.[82] For sugarcane, only five plants were acclimatized to outdoor conditions, and a number of albinos (indicative of genetic abnormalities) occurred, especially when older cultures were used.[108] With rice, one laboratory recovered only 1 plant from 1 line and 8 from another,[106] whereas another laboratory established 120 plants from a single line in soil.[100] The best plant recovery rate of over 300 rice plants was obtained using nurse culture methods.[107] Perhaps the greatest advancement in protoplast technology for the Poaceae is recently documented somatic hybridization via protoplast fusion between rice and barnyard grass (*Echinochloa oryzicola* Vasing).[109] Although all plants were abnormal and failed to mature, this technique shows promise for future combination of sexually incompatible germplasm.

The low plant recovery rates, genotype specificities, and genetic abnormalities present in even the best of circumstances severely limit practical applications of protoplast technology. However, the consistent record of recent technical advances suggests that protoplast culture will become increasingly available as a tool for germplasm manipulation.

C. Somatic Embryo Ontogeny

In vitro culture systems provide a microbiological approach to basic studies of plant development by allowing very large populations of cells to be reared and manipulated in highly controlled environments with little space and manpower expenditure. Therefore, highly complex developmental processes that normally occur only during specific stages of the plant life cycle can be studied in the test tube. Embryogenic cultures, in particular, offer a novel alternative to conventional use of seeds for studies of embryo development. Somatic embryos develop without the obscuring endosperm, nucellus, and seedcoat that are present in developing seed. Therefore, the process of embryogenesis can be more closely monitored. In addition, all embryo stages are present in a culture at a single time, so that ontogeny can be easily traced without serial sampling from mature plants, as is necessary for most studies of zygotic embryos in seeds.

Despite the amenability of embryogenic cultures for ontogenetic studies, certain aspects of somatic embryo development remain obscure and controversial. For example, somatic embryos have been variously reported to originate either from single cells[50,110-112] or by a process of budding from groups of underlying cells.[113] The purpose of this section is to trace the major events associated with somatic embryo development from early stages through germination and to discuss various interpretations of embryogenesis.

1. Early Cellular Events

An embryo can be defined as a new individual arising from a single cell and having no vascular connection with maternal tissue. Because somatic embryos often develop with a broad attachment to underlying tissue, instead of a fine connection via a suspensor, confusion has occurred as to whether or not they grow from single cells. However, examination of classical literature dealing with zygotic embryogenesis[114] has shown that a distinct developmental stage often occurs between the first cell divisions and actual embryo initiation, particularly in primitive vascular plants. In these instances, a mass of relatively disorganized embryogenic cells, termed a proembryonal complex, is formed from which embryos eventually grow. Thus, ''budding'' of somatic embryos can be interpreted as reversion to a primitive form of embryogenesis involving a proembryonal complex. This reversion appears to be a consequence of plant tissue culture. As previously discussed in Section II, formation of somatic embryos in this manner may be similar to shoot organogenesis.

In understanding the various means by which somatic embryos develop, it is necessary to consider early cell divisions in organized tissues that lead to somatic embryogenesis. In other words, what are the changes that occur during initiation of somatic embryogenesis in cells that normally would have differentiated into other cell types? In Section III.A.1, explant organs and tissues capable of somatic embryogenesis were discussed. Several studies have delineated the type and location of inducible cells within these explants. For zygotic embryo explants, the origin of embryogenic callus proliferation appears to vary from species to species. With *S. bicolor*,[31] corn,[24] and others, it appears that the scutellum gives rise to embryogenic callus. The coleorhiza[25] and scutellum[86] of orchardgrass produces embryogenic callus, whereas it is the epiblast (which may be analogous to coleorhiza) of wheat that proliferates.[115,116] The mesocotyl of *Echinochloa muricata* (Beauv.) Fern. and *E. crus-galli* (L.) Beauv. var. oryzicola (Vasing) Ohwi[91] produces embryogenic callus. Histological examination of leaf basal meristem explants have variously suggested that embryogenic cells arise from epidermis, mesophyll, and/or vascular parenchyma.[50,51,57] In all instances, highly meristematic cells are thought to be the source of the embryogenic response. Responsive cells of the scutellum may be ''embryonic'' prior to the time of culture initiation. Thus, stimulation of these cells to form new embryos may not represent a drastic change in normal development. However, leaf basal meristem cells are not destined to form embryonic tissue, and the tissue culture process clearly changes their developmental pathway.

Histology of newly cultured explants has shown that the responsive cells become densely cytoplasmic and exhibit intense mitotic activity.[110,113] Both periclinal and anticlinal divisions occur. In embryo scutellar tissue, divisions of many superficial cells lead to a characteristic folded appearance.[113] In leaf tissue, division of mesophyll cells can lead to either an embryogenic callus[50,57] or directly to embryo formation.[57] In the former instance, embryogenic callus is equivalent to a proembryonal complex and represents a deviation from the normal zygotic embryo-type developmental pattern, whereas a suspensor is formed during direct embryogenesis, which more closely approximates zygotic embryo development. Detailed histological studies of "budding" of *S. bicolor* embryos from scutellar tissue shows apparent division of single cells leading to *de novo* embryogenesis.[113] Likewise, in pearl millet, morphology of the attachment of the embryo to scutellar tissue would suggest a budding phenomena; however, histology shows that subepidermal cells divide to produce discrete cell groups that develop either directly into embryos or into embryogenic callus.[110] Here, both direct embryogenesis and proembryonal complex development occurs. Thus, although great flexibility in specific pathways of embryogenesis occurs, a single cell origin for somatic embryos, either via a proembryonal complex or directly from the explant is suggested by these studies.

2. Development of Proembryos and Embryo Maturation

Cell division patterns leading to proembryo formation vary depending upon whether the embryo is to develop with a suspensor attachment (Figure 5) or by budding with a broad basal attachment to underlying tissue (Figure 6). The exact sequence of events associated with the budding phenomena is not well understood and may be quite variable. Somatic embryogenesis, via suspensor development approximates zygotic embryogenesis and is better understood. Suspensor and proembryo formation appears to occur by a well-defined series of periclinal and anticlinal divisions. Variation in the exact number of divisions in each plane accounts for the uni- and multi-seriate suspensors that have been documented.[111] The basal cells become characteristically vacuolate[110] and are postimitotic, whereas distal cells remain densely cytoplasmic and are destined to differentiate into various embryonic tissues.

The subsequent events of embryo differentiation have been reported for orchardgrass[112] and are summarized here. Cells of globular embryos are densely cytoplasmic and appear undifferentiated from each other. During further development, embryos become elongate and clavate due to cell division and enlargement. At this stage, large vacuolate cells (containing starch) destined to become scutellar tissue as well as a region of relatively surface-oriented, densely cytoplasmic cells (indicative of nonscutellar tissues) are evident.

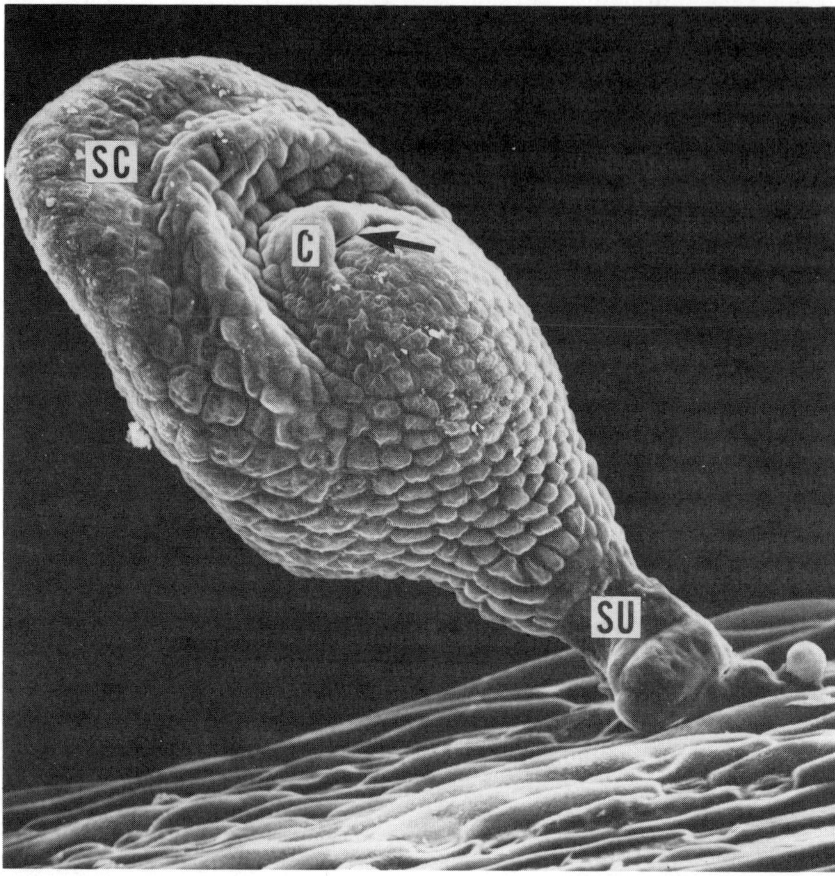

FIGURE 5. Direct embryogenesis from cultured leaf of orchardgrass. Note notch formed by coleoptile development (arrow). Coleoptile, C; scutellum, sc; suspensor, su. (From Conger, B. V., Hanning, G. E., Gray, D. J., and McDaniel, J. K., *Science*, 221, 850, 1983. Copyright 1983 by the AAAS. With permission.)

Further elongation and division of the cytoplasmic cells eventually produce a lateral notch, visible on one side of the embryo surface. This notch marks the future location of the shoot apex and represents the earliest morphological sign of coleoptile development. As the scutellum enlarges by cell division, the coleoptile eventually grows outward by similar organized divisions. During coleoptilar development, the notch initially becomes more prominent (Figure 5), but eventually closes (Figure 6) as the coleoptile enlarges to become a protective covering for the shoot apex. At this stage, scutellar cells typically contain an abundance of starch and lipids. Figures 5 and 6 show well-developed embryos prior to germination. Deeply seated meristematic cells lo-

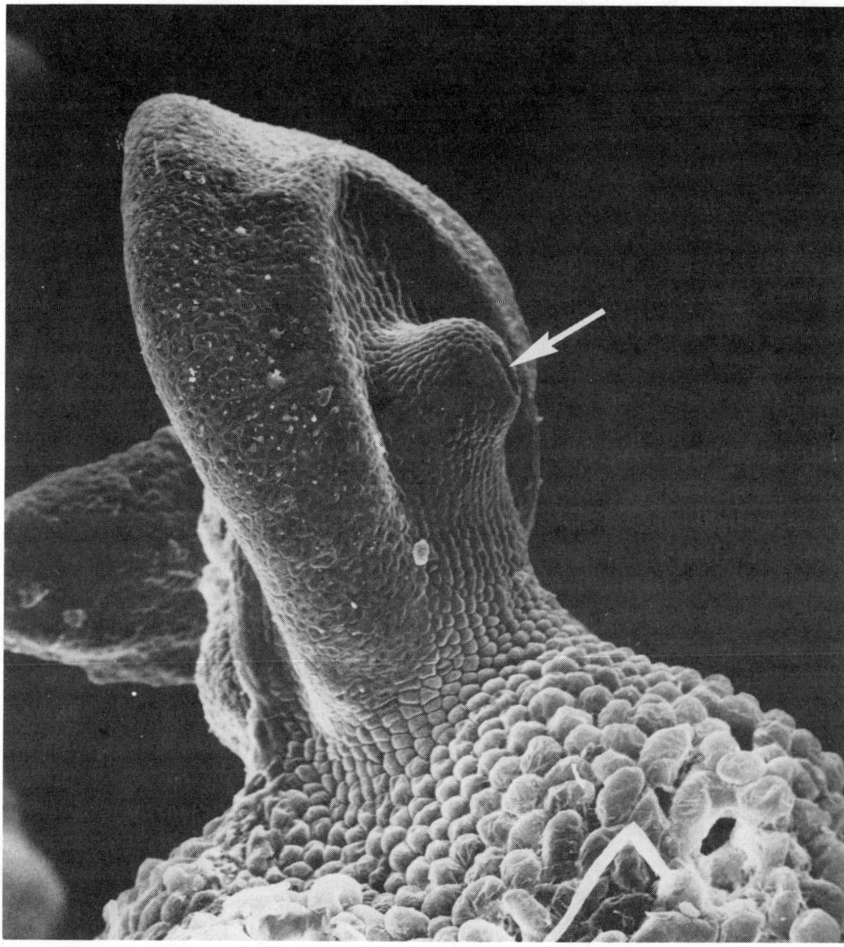

FIGURE 6. Indirect embryogenesis in orchardgrass. Note broad attachment of embryo to proembryonal complex. Further development of coleoptile (compare with Figure 5) results in closure of notch (arrow). (From Gray, D. J., Conger, B. V., and Hanning, G. E., *Protoplasma*, 122, 196, 1984. With permission.)

cated beneath the coleoptilar area divide rapidly and eventually become the root-shoot axis of the embryo. During these divisions, the scutellum becomes depleted of starch and lipids, presumably to ''feed'' development of the embryo axis. Concurrently, cells of the embryo axis become laden with lipids. Morphologically, the embryo begins to swell at the precise location of the axis due to proliferation of meristematic cells. Time-lapse studies have shown that embryo development from a late globular to a swollen mature stage capable of germination occurs in approximately 7 d.[117] For further development, embryos or cultures bearing embryos must be transferred to medium lacking growth regulators. Under this condition, the entire embryo swells and begins to assume a more disorganized appearance as the scutellum erodes. The first sign of germination is development of purplish pigmentation and the appearance of root hairs on the coleorhiza region. Shoot and root primordia begin to form. The coleoptile becomes pigmented and elongates. A root grows through and ruptures the coleorhiza region, and leaves grow through the tip of the coleoptile. Eventually, the scutellum and other remaining embryonic tissues completely erode so that only a plantlet remains.

This sequence of somatic embryo development is virtually identical to that which occurs during zygotic embryogenesis. Furthermore, the relative ease of studying somatic vs. zygotic embryogenesis allows better resolution of certain developmental events. For example, origin of the coleoptile and scutellum has been a controversial point in studies of grass embryogenesis. Both have been variously regarded to be either leaves originating from the shoot apex or embryonic organs such as rudimentary cotyledons (see Reference 118 for discussion). Studies of orchardgrass have clearly shown that both scutellum and coleoptile are initiated early in development before the root-shoot axis is differentiated.[111] Thus, these organs must be of true of embryonic origin and do not represent modified leaves.

The developmental sequence presented above represents a view of normal ontogeny; however, a number of abnormal developmental patterns also occur and actually appear to be more prevalent in somatic embryos than in their zygotic counterparts. For example, development of fused embryos is common.[111,113,117] Fused embryos occur when single proembryos abnormally develop anomalous embryonic organs. When this occurs very early in development, complete twin embryos capable of germination and supported by a single suspensor may be formed.[117] Anomolous growth occurring later in development results in single embryos with extra organs, i.e., scutellums, roots, shoots, etc. Abnormally large or odd-shaped organs develop when extra cell divisions occur only at specific developmental stages. This subject has previously been discussed in depth.[119]

A major need for refining embryogenic systems is to ''normalize'' somatic embryo development. However, efforts to manipulate embryogenic culture systems so as to increase embryo normality have been few and largely unsuccessful for grasses and cereals. Development of methods to produce well-developed, mature embryos with high germination rates will not only help resolve basic aspects of embryogenesis, but will allow practical utilization of somatic embryos as synthetic seeds, as discussed in Section IV.C.

IV. APPLICATIONS OF SOMATIC CELL CULTURE FOR CULTIVAR IMPROVEMENT

Tall fescue is an allohexaploid composed of three closely related genomes with regulator genes that confer diploid-like meiotic behavior.[120] However, tall fescue is predominantly self-sterile, presumably due to meiotic irregularities which result from reduced homeostasis in inbred populations, segregation of recessives deleterious to meiosis that are masked in the polyploid, and pairing of nonhomolgous chromosomes.[121] As with other plants, inbreeding results in smaller panicles and fewer seeds per plant as well as reduced vigor and forage quality.[122] Therefore, true-breeding seed is not desirable for stand establishment, but is useful in hybrid and cultivar development.

Because tall fescue lacks organs suitable for mass vegetative propagation, such as stolons and rhizomes, seed is the only economical means of planting. Several open-pollinated lines have been selected and released as cultivars; however, directed improvement through breeding has relied on development of synthetic strains. Development of synthetic cultivars is exceedingly time consuming, since up to 43 clone lines must be selected which will produce vigorous, high-quality, and phenotypically uniform plants from cross-pollinated seed.[123] Because of the difficulty associated with synthetic cultivar development, it is very time consuming to incorporate desirable, new traits into an existing tall fescue cultivar and impossible to produce true seed from a single outstanding selection.

Techniques of somatic cell culture offer potential alternatives to these conventional approaches by (1) allowing specific traits to be selected from or introduced into otherwise desirable cultivars without breeding, or (2) providing efficient vegetative propagation systems as an alternative to planting crops via seed. This section will explain applications of specific cell culture techniques that may prove useful for tall fescue improvement.

A. Selection of Agronomic Traits

Advances in methods to grow large populations of totipotent cells *in vitro* have increasingly allowed microbial technology to be applied to plant genetic manipulation. Preexisting or induced variation can be manipulated in such large populations. Variation arising as a consequence of cell culture has been termed somaclonal variation.[124] Any variation in cell cultures can potentially be selected by two general means: (1) *in vitro* selection is possible when a selection agent that can be added to cell cultures is available (variant cells able to grow in the presence of the selection agent persist, and plants regenerated from such cells should possess the selected characteristic), and (2) plants may be regenerated from nonselected cultures and later evaluated for variation that presumably existed either in the explant prior to culture or arose via the tissue culture process. Thus, *in vitro* selection allows directed selection of desirable traits, whereas postcultural evaluation of regenerants represents a broader approach and requires a relatively time-consuming analysis of many plants. Variation selected by these means may or may not result from mutation. However, heritability of selected traits has been documented,[125-127] thus demonstrating that selection of mutants is possible by these means. For seed-propagated crops, selected traits must be heritable. Likewise, heritability is needed if selected plants are to be used in breeding. Selection of stable, nonheritable traits is useful only if the resulting plants are of improved cultivar quality and can be clonally propagated.

Tall fescue may be a good candidate for these selection methods because major variation has been observed in plants regenerated from tissue culture.[1,5,128] If a large degree of variation occurs in tall fescue tissue culture, it is possible that useful traits can be accessed through selection.

1. In Vitro Selection

In vitro selection is possible only where variation or traits desired in the whole plant are operative at the single-cell level. The desired trait or variation should exist in natural populations so as to provide confidence that selection is possible. Selection agents operate by killing of inhibiting growth of nondesirable cells; thus, desired traits can be obtained only in instances where such selection agents are available. Because of this, *in vitro* selection agents and selectable traits are limited. Possible selection agents and selectable traits are (1) specific toxins or culture filtrates from disease-causing organisms for selecting disease resistance, (2) herbicides for selecting herbicide resistance, and (3) salts (e.g., NaCl) and metals (e.g., aluminum) for selecting environmental stress tolerance. Typically, the selection agent is added directly to culture medium at a critically adjusted concentration. The best chance of selecting an existing cell with a heritable trait is to apply a lethal dose of selection agent. The alternative approach of gradually increasing sublethal dosages may result in physiologically acclimated cell lines without heritable resistance.

Several successful examples of *in vitro* selection exist among poaceous species. The best examples are those dealing with disease resistance. Using toxin selection methodology, maternally inherited resistance to race T of *Helminthosporium maydis* Nisik. & Miyake was selected in corn,[129,130] and heritable resistance of sugarcane to *H. sacchari* (B. de Haan) Butl. was selected as well.[131] Recently, selection with bacterial blight toxin yielded several wheat plants with resistance to the causal organism.[132] The majority of diseases of tall fescue are caused by fungi.[133] Certain of these organisms induce symptoms reminiscent of those caused by a toxin. For example, *H. dictyoides* Dresch. produces dark brown streaks of tissue in lesions on leaves. *Rhizoctonia solani* Kuhn causes a rapid leaf scald. *Cercospora apii* Fes. causes coalescing lesions with purple-brown margins.[133] Perhaps toxin selection could be used in applicable cases to enhance resistance in tall fescue.

2. Selection of Regenerated Plants

Selection for desirable traits in regenerated plants is dependent on the assumption that frequent genetic changes occur during the tissue culture process.[124] The advantage of field selection over *in vitro* selection is that selectable traits are not limited to available selection agents. Thus, not only is selection for resistance to various environmental stresses possible by this approach, but evaluation for complex multigenic traits can be considered. Additionally, specific morphotypes such as improved turf or forage varieties may arise among regenerants.

Several examples of selection from regenerated poaceous plants exist. For sugarcane, resistances to Fiji disease, *Sclerospora sacchari* Miy., and *H. sacchari* were selected, although heritability was documented only in the former.[134] Culture-induced mutations in corn were similar to known spontaneous mutations; however, significant qualitative and quantitative variation in morphology and growth dynamics were found.[135] Heritable variation in morphological characteristics that did not occur in control seed-derived plants was found in tissue culture-derived wheat plants.[136]

Some variation arising from tissue culture may be detrimental, however, and may spoil use of regenerated plants as parental lines. For example, one breeding study of regenerated wheat plants showed that most transmitted variation was negative, including partial sterility and a number of chromosonal abnormalities.[137] It is also important to consider alternatives in evaluating variations among regenerated plants of heterozygous species such as forage grasses. It is yet unclear as to whether or not more variation exists in regenerated plants than in seed progeny. For example, salt-tolerant plants of orchardgrass, red fescue (*Festuca rubra* L.), ryegrass, and *Holeus lanatus* L. were selected from seed-derived plants of open-pollinated germplasm grown on sand with elevated NaCl levels.[138] Likewise, multiresistance to three rust diseases was selected in wheat plants grown from seeds mutagenized with nitrosomethyl urea.[139] This suggests that there is a high degree of useful variation in existing populations that can be readily selected without the added expense of cell culture. Thus, especially in heterozygous species, variation accessed by conventional breeding may be a more lucrative approach than selection of regenerated plants, since potential problems of major chromosomal aberrancies may be reduced.

B. Introduction of Agronomically Useful Genes

Although progress in genetic engineering of the Poaceae has lagged behind that of other major plant families, recent advances in somatic hybridization and molecular genetics technology provide optimism that rapid introduction of new and improved agronomic traits into select plants will become a future reality. Our ability to regenerate certain cereal and grass plants from protoplasts, as described in Section III.B.3, allows development of somatic hybridization technology using protoplast fusion and facilitates certain aspects of genetic engineering. This section describes new advances in somatic cell genetics of the Poaceae.

1. Somatic Hybridization

Somatic hybridization via protoplast fusion is potentially useful for combining germplasm from sexually incompatible species. A major problem with this approach, however, is that the resulting somatic hybrid contains twice as many chromosomes as either "parental". This limits use of somatic hybrids except in instances where (1) the somatic hybrid is in itself commercially useful, or (2) where breeding programs utilizing the resulting increased ploidy level exist. Potential alternative solutions to the problem of increased ploidy level are (1) to fuse protoplasts of gametic or haploid cells so that the resultant hybrid exists at the normal somatic ploidy level, or (2) to fuse nucleated protoplasts of one species with enucleated protoplasts of another species so that the resulting cells (termed cybrids) contain hybrid cytoplasm, but only the base nuclear chromosome number of one parent. Neither of these techniques has yet been developed for the Poaceae.

Tangible progress in somatic hybridization of the Poaceae has occurred only recently. Hybrid cell lines have been produced by protoplast fusion of somatic cells of sugarcane + pearl millet,[140] and pearl millet + guinea grass.[141] Although plants were not obtained, the hybrid nature of resulting colonies was confirmed by isozyme analysis and somatic embryo-like structures were documented for the former combination. Embryogenesis and plant regeneration were obtained by protoplast fusion of rice and barnyard grass;[109] however, plants were highly abnormal and failed to mature. This represents the farthest development of somatic hybridization technology for grasses and cereals at this time (1989). However, recent advances in plant regeneration from protoplasts of tall fescue as well as other forage species[4] strongly suggests that somatic hybridization technology will soon be available for these grasses.

2. *Genetic Transformation*

Specific and directed manipulation of plant phenotype is becoming available by techniques of molecular genetics. The insertion of discrete genes into plant cells has been accomplished in model dicot systems, and in some instances, expression and inheritance of the genes have been obtained.[142] All well-defined transformation systems rely on use of the Ti plasmid contained in *Agrobacterium tumefaciens* as a means of gene insertion. *Agrobacterium* is not generally regarded to be infective in monocots, and this has severely limited progress in genetic transformation in the Poaceae. However, recent evidence does suggest that *Agrobacterium* can deliver DNA from maize streak virus into maize plants.[143] Thus, it is possible that *Agrobacterium* has a wider host range than first perceived. Application of *Agrobacterium* to transformation in grasses and cereals will probably depend on refining efficient Ti plasmid promoter sequences for monocots.

Other methods of direct gene insertion probably offer more immediate prospects for producing transgenic grasses and cereals. For example, a derivative Ti plasmid was injected directly into young floral tillers of rye, and a marker gene was subsequently expressed in developing seeds, thus showing transformation.[144] This approach circumvents cell culture completely. Insertion of DNA into cells, protoplasts, or pollen grains by microinjection or newly developed particle gun technology may also prove successful. The current state of the art in genetic engineering of cereals, however, employs electroporation of DNA into protoplasts. An electric current pulse causes the plasmalemma to temporarily solubilize, and surrounding DNA particles are engulfed. This technique was demonstrated with oat[145] and has yielded transgenic corn plants.[146] A marker gene was expressed in the transformed corn tissue. Unfortunately all regenerated plants were sterile, so that heritability could not be assessed. However, successful transformation of corn demonstrates utility of basic techniques of molecular genetics for the Poaceae. Successful regeneration of tall fescue from protoplasts[4] will now allow application and evaluation of such techniques. Perhaps the greatest limitation to utilizing molecular genetics for crop improvement is the current lack of agronomically useful genes. Thus far, the most immediate prospects for producing improved cultivars is by the insertion of genes that have been isolated for disease and herbicide resistance. For example, bacterial genes which code for proteins toxic to certain insects have been inserted into tobacco plants.[147] The transgenic plants became toxic to feeding larvae. The inventory of useful genes and gene combinations available to agronomists will doubtlessly grow during the near future.

C. *In Vitro* Mass Cloning

A major barrier to rapid improvement of self-incompatible, seed-propagated crops like tall fescue is the difficulty associated with producing true seed from outstanding selections. The ability to vegetatively propagate at commercially acceptable levels, as is possible with stoloniferous species like bermudagrass, would alleviate the need for seed. However, tall fescue does not possess suitable structures, such as rhizomes and stolons, that are amenable to vegetative propagation. Although tall fescue plant number can be easily increased by the separation and replanting of tillers, this mode of propagation is too slow and expensive for stand establishment.

Methods of tissue culture propagation hold potential for increasing multiplication rates to acceptable efficiencies. However, because of the low per-plant value of tall fescue, propagation costs must approach that of seed. Plant regeneration from callus culture is possible (see Section I), but is currently too inefficient for commercial use. The best prospect of cloning at seed efficiencies is by use of somatic embryos.[148] Somatic embryogenesis offers the greatest efficiencies in production and clonal plant recovery that can be envisioned (see Sections II and III). Somatic embryos are nearly identical to zygotic embryos, and this suggests that they may be manipulated or modified to possess the efficient handling quality of seeds. Such "synthetic seeds" would be as easy to utilize as natural seed, but would give rise to clonal plants. However, more research and development is needed before fragile, rapidly growing somatic embryos can be engineered to mimic durable, quiescent, easily manipulated seed.[146]

Several reports have discussed the development and applications of synthetic seed technology.[148,149] For synthetic seed to become useful, technical advances must occur in the areas of embryo production, maturation, and quiescence, as well as in the development of artificial seed coatings. Feasibility at a prototype level has been demonstrated for each of these areas.[148-152] Thus far, it is possible to produce large numbers of somatic embryos of certain species in liquid suspension cultures (see Section III.B.2), induce a quiescent phase by dehydration,[152] and, for nondehydrated embryos, a protective coating has been developed.[149,151] The state of the art for synthetic seeds of poaceous species exists for orchardgrass, where somatic embryos dehydrated to 13% water were stored in a quiescent state at room temperature for 21 d; 4% of the embryos germinated into plants after imbibition.[153] Although, these low plant recovery rates do not hold practical application, the feasibility of causing somatic embryos to behave as seeds with respect to a stable resting stage is clearly indicated.

The driving force behind development of synthetic seeds for crops like tall fescue is the possibility of circumventing the problems associated with self-incompatibility for cultivar improvement. Improved genotypes produced by either conventional hybridization or by any of the means discussed in Sections IV A and B could be commercially propagated without development of multilines. Successful application of synthetic seed technology would thus revolutionize cultivar improvement of tall fescue. Potential drawbacks of using synthetic seed technology for tall fescue improvement lie in the cost of research for systems development and the expected cost differential between production of normal seed and synthetic seed. The relative low value of forage grasses when compared to major agronomic crops will translate into less emphasis on such extended developmental research. However, it is anticipated that basic advances in technology developed for major cereals will be applicable to the forages, thus reducing developmental expenses.

Somatic cell culture methodology for improvement of forage grasses is still at an early stage of development. However, the anticipation of so many significant applications to cultivar improvement is creating much research activity worldwide. The future of plant improvement holds great challenges and much excitement as various somatic cell systems are refined and put to use.

REFERENCES

1. **Lowe, K. W. and Conger, B. V.,** Root and shoot formation from callus cultures of tall fescue, *Crop Sci.,* 19, 397, 1979.
2. **Kasperbauer, M. J., Buckner, R. C., and Springer, W. D.,** Haploid plants by anther-panicle culture of tall fescue, *Crop Sci.,* 20, 103, 1980.

3. **Dale, P. J. and Dalton, S. J.,** Immature inflorescence culture in *Lolium, Festuca, Phleum* and *Dactylis, Z. Pflanzenphysiol.,* 111, 39, 1983.
4. **Dalton, S. J.,** Plant regeneration from cell suspension protoplasts of *Festuca arundinacea* Schreb., *Lolium perenne* L., and *L. multiflorum* Lam., *Plant Cell Tissue Organ Cult.,* 12, 137, 1988.
5. **Kasperbauer, M. J., Buckner, R. C., and Bush, L. P.,** Tissue culture of annual ryegrass × tall fescue hybrids: callus establishment and plant regeneration, *Crop Sci.,* 19, 457, 1979.
6. **Murashige, T. and Skoog, F.,** A revised medium for rapid growth and bioassays with tobacco cultures, *Physiol. Plant.* 15, 483, 1962.
7. **Chen, C. H., Chen, L. F., Lo, P. F., and Ross, J. G.,** Plant regeneration from cultured immature inflorescences of orchardgrass (*Dactylis glomerata* L.), *Euphytica,* 31, 19, 1982.
8. **Chen, C. H., Songstad, D. D., and Boe, A. A.,** Direct plant regneration from inflorescence cultures of orchardgrass (*Dactylis gomerata* L.), *Proc. S. D. Acad. Sci.,* 64, 77, 1985.
9. **Conger, B. V. and McDonnell, R. E.,** Plantlet formation from cultured inflroescences of *Dactylis glomerata* L., *Plant Cell Tissue Organ Cult.,* 2, 191, 1983.
10. **Conger, B. V. and Carabia, J. V.,** Callus induction and plantlet regeneration in orchardgrass, *Crop Sci.,* 18, 157, 1978.
11. **Eapen, S. and Rao, P. S.,** Plant regeneration from callus cultures of *Durum* and Emmer wheat, *Plant Cell Rep.,* 1, 215, 1982.
12. **Eapen, S. and Rao, P. S.,** Organogenesis and plantlet formation from callus cultures of different cultivars of bread wheat *(Triticum aestivum), Proc. Indian Natl. Sci. Acad.,* 48, 371, 1982.
13. **Green, C. E. and Phillips, R. L.,** Plant regeneration from tissue cultures of maize, *Crop Sci.,* 15, 417, 1975.
14. **Heinz, D. J. and Mee, G. W. P.,** Plant differentiation from callus tissue of *Saccharum* species, *Crop Sci.,* 9, 346, 1969.
15. **Krans, J. V., Henning, V. T., and Torres, K. C.,** Callus induction, maintenance and plantlet regeneration in creeping bentgrass, *Crop Sci.,* 22, 1193, 1982.
16. **Ling, D. H., Chen, W. Y., Chen, M. F., and Ma, Z. R.,** Direct development of plantlets from immature panicles of rice in vitro, *Plant Cell Rep.,* 2, 172, 1983.
17. **Luppotto, E.,** Callus induction and plant regeneration from barley mature embryos, *Ann. Bot.,* 54, 523, 1984.
18. **McDonnell, R. E. and Conger, B. V.,** Callus induction and plantlet formation from mature embryo explants of Kentucky bluegrass, *Crop Sci.,* 24, 573, 1984.
19. **Nabors, M. W., Kroskey, C. S., and McHugh, D. M.,** Green spots are predictors of high callus growth rates and shoot formation in normal and in salt stressed tissue cultures of oat (*Avena sativs* L.), *Z. Pflanzenphysiol.,* 105, 341, 1982.
20. **Nakano, H. and Maeda, E.,** Morphology of the process of shoot formation in the rice callus culture, *Proc. Crop Sci. Soc. Jpn.,* 43, 151, 1974.
21. **Skene, K. G. M. and Barlass, M.,** Regeneration of plants from callus cultures of *Lolium rigidum, Z. Pflanzenzuecht.,* 90, 130, 1983.
22. **Yan, C. J. and Zhao, Q. H.,** Callus induction and plantlet regeneration from leaf blade of *Oryza sativa* L. subsp. indica, *Plant Sci. Lett.,* 25, 187, 1982.
23. **Zamora, A. B. and Scott, K. J.,** Callus formation and plant regeneration from wheat leaves, *Plant Sci. Lett.,* 29, 183, 1983.
24. **Green, C. E. and Armstrong, C. L.,** Somatic cell genetic systems in corn, in *Advances in Gene Technology: Molecular Genetics of Plants and Animals,* (Miami Winter Symp. Ser.), Vol. 20, Fazelahmad, A., Downey, K., Schultz, J., Voellmy, R. W., Eds., Academic Press, New York, 1983, 1.

25. **McDaniel J. K., Conger, B. V., and Graham, E. T.,** A histological study of tissue proliferation, embryogenesis, and organogenesis from tissue cultures of *Dactylis gomerata* L., *Protoplasma,* 110, 121, 1982.
26. **Zimny, J. and Lorz, H.,** Plant regeneration and initiation of cell suspensions from root-tip derived callus of *Oryza sativa* L. (rice), *Plant Cell Rep.,* 5, 89, 1986.
27. **Lowe, K., Taylor, D. B., Ryan, P., and Paterson, K. E.,** Plant regeneration via organogenesis and embryogenesis in the maize inbred line B73, *Plant Sci.,* 41, 125, 1985.
28. **Williams, E. G. and Maheswaran, G.,** Somatic embryogenesis: factors influencing coordinated behaviour of cells as an embryogenic group, *Ann. Bot.,* 57, 443, 1986.
29. **Broertjes, C. and Van Harten, A. M.,** Single cell origin of adventitious buds, *Euphytica,* 34, 93, 1985.
30. **Horsch, R. B., Fry, J. E., Hoffman, N. L., Eichholtz, D., Robergs, S. G., and Fraley, R. J.,** A simple method for transferring genes into plants, *Science,* 227, 1229, 1985.
31. **Thomas, E., King, P. J., and Potrykus, I.,** Shoot and embryo-like structure formation from cultured tissues of *Sorghum bicolor, Naturwissenschaften,* 64, 587, 1977.
32. **Norstog, K. and Smith, J. E.,** Culture of small barley embryos on defined media, *Science,* 142, 1655, 1963.
33. **Dale, P. J.,** Embryoids from cultured immature embryos of *Lolium multiflorum, Z. Pflanzenphysiol.,* 100, 73, 1980.
34. **Eapen, S. and Rao, P. S.,** Callus induction and plant regeneration from immature embryos of rye and triticale, *Plant Cell Tissue Organ Cult.,* 1, 221, 1982.
35. **Heyser, J. W., Dykes, T. A., DeMott, K. J., and Nabors, M. W.,** High frequency, long term regeneration of rice from callus culture, *Plant Sci. Lett.,* 29, 175, 1983.
36. **Heyser, J. W. and Nabors, M. W.,** Regeneration of proso millet from embryogenic calli derived from various plant parts, *Crop Sci.,* 22, 1070, 1982.
37. **Heyser, J. W. and Nabors, M. W.,** Long term plant regeneration, somatic embryogenesis and green spot formation in secondary oat *(Avena sativa)* callus, *Z. Pflanzenphysiol.,* 107, 153, 1982.
38. **Lu, C., Vasil, I. K., and Ozias-Akins, P.,** Somatic embryogenesis in *Zea mays* L., *Theor. Appl. Genet.,* 62, 109, 1982.
39. **Lu, C. Y. and Vasil, I. K.,** Somatic embryogenesis and plant regeneration from freeley-suspended cells and cell groups of *Panicum maximum* Jacq., *Ann. Bot.,* 48, 543, 1981.
40. **Ozias-Akins, P. and Vasil, I. K.,** Plant regeneration from cultured immature embryos and inflorescences of *Triticum aestivum* L. (wheat): evidence for somatic embryogenesis, *Protoplasma,* 110, 95, 1982.
41. **Vasil, V. and Vasil, I. K.,** Isolation and culture of cereal protoplasts, *Theor. Appl. Genet.,* 56, 97, 1980.
42. **Heyser, J. W., Nabors, M. W., Mackinnon, C., Dykes, T. A., DeMott, K. J., Kautzman, D. C., and Kazi-Mujeeb, A.,** Long-term, high-frequency plant regeneration and the induction of somatic embryogenesis in callus cultures of wheat (*Triticum aestivum* L.), *Z. Pflanzenzuecht.,* 94, 218, 1985.
43. **Vasil, V., Lu, C. Y., and Vasil, I. K.,** Histology of somatic embryogenesis in cultured immature embryos of maize (*Zea mays* L.), *Protoplasma,* 127, 1, 1985.
44. **Botti, C. and Vasil, I. K.,** Plant regeneration by somatic embryogenesis from parts of cultured mature embryos of *Pennisetum americanum* (L.) K. Schum, *Z. Pflanzenphysiol.,* 111, 319, 1983.
45. **Blanche, F. C., Krans, J. V., and Coats, G. E.,** Improvement in callus growth and plantlet formation in creeping bentgrass, *Crop Sci.,* 26, 1245, 1986.
46. **Nabors, M. W., Heyser, J. W., Dykes, T. A., and Demott, K. J.,** Long-duration, high-frequency plant regeneration from cereal tissue cultures, *Planta,* 157, 385, 1983.
47. **Torello, W. A., Symington, A. G., and Rufner, R.,** Callus initiation, plant regeneration, and evidence of somatic embryogenesis in red fescue, *Crop Sci.,* 24, 1037, 1984.

48. **MacKinnon, C., Gunderson, G., and Nabors, M. W.,** High efficiency plant regeneration by somatic embryogenesis from callus of mature embryo explants of bread wheat *(Triticum aestivum)* and grain sorghum *(Sorghum bicolor), In Vitro Cell. Dev. Biol.,* 23, 443, 1987.
49. **Chang, Y. F.,** Plant regeneration in vitro from leaf tissues derived from cultured immature embryos of *Zea mays* L., *Plant Cell Rep.,* 2, 183, 1983.
50. **Ho, W. J. and Vasil, I. K.,** Somatic embryogenesis in sugarcane *(Saccharum officinarum* L.) I. The morphology and physiology of callus formation and the ontogeny of somatic embryos, *Protoplasma,* 118, 169, 1983.
51. **Lu, C. Y. and Vasil, I. K.,** Somatic embryogenesis and plant regeneration from leaf tissues of *Panicum maximum* Jacq., *Theor. Appl. Genet.,* 59, 275, 1981.
52. **Wernicke, W. and Brettell, R.,** Somatic embryogenesis from *Sorghum bicolor* leaves, *Nature (London),* 287, 138, 1980.
53. **Wernicke, W., Brettell, R., Wakizuka, T., and Potrykus, I.,** Adventitious embryoid and root formation from rice leaves, *Z. Pflanzenphysiol.,* 103, 361, 1981.
54. **Wernicke, W. and Milkovits, L.,** Developmental gradients in wheat leaves — response of leaf segments in different genotypes cultured *in vitro, J. Plant Physiol.,* 115, 49, 1984.
55. **Hanning, G. E. and Conger, B. V.,** Embryoid and plantlet formation from leaf segments of *Dactylis glomerata* L., *Theor. Appl. Genet.,* 63, 155, 1982.
56. **Haydu, A. and Vasil, I. K.,** Somatic embryogenesis and plant regeneration from leaf tissues and anthers of *Pennisetum purpureum* Schum, *Theor. Appl. Genet.,* 59, 269, 1981.
57. **Conger, B. V., Hanning, G. E., Gray, D. J., and McDaniel, J. K.,** Direct embryogenesis from mesophyll cells of orchardgrass, *Science,* 221, 850, 1983.
58. **Conger, B. V., Novak, F. J., Afza, R., and Erdelsky, K.,** Somatic embryogenesis from cultured leaf segments of *Zea mays, Plant Cell Rep.,* 6, 345, 1987.
59. **Linacero, R. and Vazquez, A. M.,** Somatic embryogenesis and plant regeneration from leaf tissues of rye (*Secale cereale* L.), *Plant Sci.,* 44, 219, 1986.
60. **Hanning, G. E. and Conger, B. V.,** Factors influencing somatic embryogenesis from cultured leaf segments of *Dactylis glomerata, J. Plant Physiol.,* 123, 23, 1986.
61. **Wernicke, W., Potrykus, I., and Thomas, E.,** Morphogenesis from cultured leaf tissue of *Sorghum bicolor* — the morphogenetic pathways, *Protoplasma,* 111, 53, 1982.
62. **Dale, P. J., Thomas, E., Brettell, R. I. S., and Wernicke, W.,** Embryogenesis from cultured immature inflorescences and nodes of *Lolium multiflorum, Plant Cell Tissue Organ Cult.,* 1, 47, 1981.
63. **Rangan, T. S. and Vasil, I. K.,** Somatic embryogenesis and plant regeneration in tissue cultures of *Panicum miliaceum* L. and *Panicum miliare* Lamk, *Z. Pflanzenphysiol.,* 109, 49, 1983.
64. **Vasil, V. and Vasil, I. K.,** Characterization of an embryogenic cell suspension culture derived from cultured inflorescences of *Pennisetum americanum* (pearl millet, Gramineae), *Am. J. Bot.,* 69, 1441, 1982.
65. **Wang, D. Y. and Vasil, I. K.,** Somatic embryogenesis and plant regeneration from inflorescence segments of *Pennisetum purpureum* Schum. (napier or elephant grass), *Plant Sci. Lett.,* 25, 147, 1982.
66. **Ahn, B. J., Huang, F. H., and King, J. W.,** Regeneration of bermudagrass cultivars and evidence of somatic embryogenesis, *Crop Sci.,* 27, 594, 1987.
67. **Yeh, M. L. and Chang, W. C.,** Somatic embryogenesis and subsequent plant regeneration from inflorescence callus of *Bambusa beecheyana* munro var. beecheyana, *Plant Cell Rep.,* 5, 409, 1986.
68. **Yeh, M. L. and Chang, W. C.,** Plant regeneration through somatic embryogenesis in callus culture of green bamboo (*Bambusa oldhamii* munro), *Theor. Appl. Genet.,* 73, 161, 1986.

69. **Zhi-hong, X., Da-yuan, W., Li-jun, Y., and Zhi-ming, W.,** Somatic embryogenesis and plant regeneration in cultured immature inflorescences of *Setaria italica, Plant Cell Rep.*, 3, 149, 1984.
70. **Ling, D. H., Chen, W. Y., Chen, M. F., and Ma, Z. R.,** Somatic embryogenesis and plant regeneration in an interspecific hybrid of *Oryza, Plant Cell Rep.*, 2, 169, 1983.
71. **Bhaskaran, S. and Smith, R. H.,** Enhanced somatic embryogenesis in *Sorghum bicolor* from shoot tip culture, *In Vitro Cell. Dev. Biol.*, 24, 65, 1988.
72. **Songstad, D. D. and Conger, B. V.,** Direct embryogenesis from cultured anthers and pistils of *Dactylis glomerata, Am. J. Bot.*, 73, 989, 1986.
73. **Schenk, R. U. and Hildebrandt, A. C.,** Medium and techniques for induction and growth of monocotyledonous and dicotyledonous plant cell cultures, *Can. J. Bot.*, 50, 199, 1972.
74. **Gamborg, D. L., Miller, R. A., and Ojima, K.,** Nutrient requirements of suspension cultures of soybean root cells, *Exp. Cell Res.*, 50, 151, 1968.
75. **Chih-ching, C., Ching-chu, W., Ching-san, S., Chen, H., Kwang-chu, Y., Chih-yin, C., and Feng-yun, B.,** Establishment of an efficient medium for anther culture of rice through comparative experiments on the nitrogen sources, *Sci. Sin.*, 18, 659, 1975.
76. **Conger, B. V., Ed.,** Agronomic crops, in *Cloning Agricultural Plants Via In Vitro Tehcniques*, CRC Press, Boca Raton, FL, 1981, 165.
77. **Papenfuss, J. M. and Carman, J. G.,** Enhanced regeneration from wheat callus cultures using dicamba and kinetin, *Crop Sci.*, 27, 588, 1987.
78. **Linsmaier, E. M. and Skoog, F.,** Organic growth factor requirements of tobacco tissue cultures, *Physiol. Plant.*, 18, 100, 1965.
79. **Green, C. E., Philips, R. L., and Kleese, R. A.,** Tissue cultures of maize (*Zea mays* L.) initiation, maintenance, and organic growth factors, *Crop Sci.*, 14, 54, 1974.
80. **Kamo, K. K., Becwar, M. R., and Hodges, T. K.,** Regeneration of *Zea mays* L. from embryogenic callus, *Bot. Gaz.*, 146, 327, 1985.
81. **Vasil, V. and Vasil, I. K.,** Plant regeneration from friable embryogenic callus and cell suspension cultures of *Zea mays* L., *J. Plant Physiol.*, 124, 399, 1986.
82. **Rhodes, C. A., Lowe, K. S., and Ruby, K. L.,** Plant regeneration from protoplasts isolated from embryogenic maize cell cultures, *Bio/Technology*, 6, 56, 1988.
83. **Duncan, D. R., Williams, M. E., Zehr, B. E., and Widholm, J. M.,** The production of callus capable of plant regeneration from immature embryos of numerous *Zea mays* genotypes, *Planta*, 165, 322, 1985.
84. **Trigiano, R. N. and Conger, B. V.,** Regulation of growth and somatic embryogenesis by proline and serine in suspension cultures of *Dactylis glomerata, J. Plant Physiol.*, 130, 49, 1987.
85. **Siriwardana, S. and Nabors, M. W.,** Tryptophan enhancement of somatic embryogenesis in rice, *Plant Physiol.*, 73, 142, 1983.
86. **Gray, D. J., Conger, B. V., and Hanning, G. E.,** Somatic embryogenesis in suspension and suspension-derived callus cultures of *Dactylis glomerata, Protoplasma*, 122, 196, 1984.
87. **Gray, D. J. and Conger, B. V.,** Influence of dicamba and casein hydrolysate on somatic embryo number of culture quality in cell suspensions of *Dactylis glomerata* (Gramineae), *Plant Cell Tissue Organ Cult.*, 4, 123, 1985.
88. **Powell, W. and Dunwell, J. M.,** *In vitro* genetics of barley (*Hordeum vulgare* L.): response of immature embryos to 2,4-dichlorophenoxyacetic acid, *Heredity*, 58, 75, 1987.
89. **Maddock, S. E., Lancaster, V. A., Risiott, R., and Franklin, J.,** Plant regeneration from cultured immature embryos and inflorescences of 25 cultivars of wheat *(Tritium aestivum), J. Exp. Bot.*, 34, 915, 1983.
90. **Vasil, V. and Vasil, I. K.,** Somatic embryogenesis and plant regeneration from suspension cultures of pearl millet *(Pennisetum americanum), Ann. Bot.*, 47, 669, 1981.

91. **Cobb, B. G., Vanderzee, D., Loescher, W. H., and Kennedy, R. A.,** Evidence of plantlet regeneration via somatic embryogenesis in the grasses *Echinochloa muricata* and *E. crus-galli* var. *oryzicola, Plant Sci.,* 40, 121, 1985.
92. **Torello, W. A., Rufner, R., and Symington, A. G.,** The ontogeny of somatic embryos from long-term callus cultures of red fescue, *HortScience,* 20, 938, 1985.
93. **Cure, W. W. and Mott, R. L.,** A comparative anatomical study of organogenesis in cultured tissues of maize, wheat and oats, *Physiol. Plant,* 42, 91, 1978.
94. **King, P. J., Potrykus, I., and Thomas, E.,** *In vitro* genetics of cereals problems and perspectives, *Physiol. Veg.,* 16, 381, 1978.
95. **Kamo, K. K. and Hodges, T. K.,** Establishment and characterization of long-term embryogenic maize callus and cell suspension cultures, *Plant Sci.,* 45, 111, 1986.
96. **Karlsson, S. B. and Vasil, I. K.,** Morphology and ultrastructure of embryogenic cell suspension cultures of *Panicum maximum* (guinea grass) and *Pennisetum purpureum* (napier grass), *Am. J. Bot.,* 73, 894, 1986.
97. **Thompson, J. A., Abdullah, R., and Cocking, E. C.,** Protoplast culture of rice (*Oryza sativa* L.) using media solidified with agarose, *Plant Sci.,* 47, 123, 1986.
98. **Ho, W. J. and Vasil, I. K.,** Somatic embryogenesis in sugarcane (*Saccharum officinarum* L.) growth and plant regeneration from embryogenic cell suspension cultures, *Ann. Bot.,* 51, 719, 1983.
99. **Jones, M. G. K. and Dale, P. J.,** Reproducible regeneration of callus from suspension culture protoplasts of the grass *Lolium multifluorum, Z. Pflanzenphysiol.,* 105, 267, 1982.
100. **Abdullah, R., Cocking, E. C., and Thompson, S. A.,** Efficient plant regeneration from rice protoplasts through somatic embryogenesis, *Bio/Technology,* 4, 1087, 1986.
101. **Kao, K. N., Gamborg, O. L., Michayluk, M. R., Keller, W. A., and Miller, R. A.,** The effects of sugars and inorganic salts on cell regeneration and sustained division in plant protoplasts, *Colloq. Int. C.N.R.S.,* 212, 207, 1973.
102. **Vasil, V. and Vasil, I. K.,** Isolation and culture of cereal protoplasts I. Callus formation from pearl millet *(Pennisetum americanum)* protoplasts, *Z. Pflanzenphysiol.,* 92, 379, 1979.
103. **Lu, C. Y., Vasil, V., and Vasil, I. K.,** Isolation and culture of protoplasts of *Panicum maximum* Jacq. (guinea grass): somatic embryogenesis and plantlet formation, *Z. Pflanzenphysiol.,* 104, 311, 1981.
104. **Vasil, V. and Vasil, I. K.,** Formation of callus and somatic embryos from protoplasts of a commercial hybrid of maize (*Zea mays* L.), *Theor. Appl. Genet.,* 73, 793, 1987.
105. **Coulibaly, M. Y. and Demarly, Y.,** Regeneration of plantlets from protoplasts of rice, *Oryza sativa* L., *Z. Pflanzenzuecht.,* 96, 79, 1986.
106. **Yamada, Y., Yang, Z-.Q., and Tang, D-.T.,** Plant regeneration from protoplast-derived callus of rice (*Oryza sativa* L.), *Plant Cell Rep.,* 5, 85, 1986.
107. **Kyozuka, J., Hayashi, Y., and Shimamoto, K.,** High frequency plant regeneration from rice protoplasts by novel nurse culture methods, *Mol. Gen. Genet.,* 206, 408, 1987.
108. **Srinivasan, C. and Vasil, I. K.,** Plant regeneration from protoplasts of sugarcane (*Saccharum officinarum* L.), *J. Plant Physiol.,* 126, 41, 1986.
109. **Terada, R., Kyozuka, J., Nishibayashi, S., and Shimamoto, K.,** Plantlet regeneration from somatic hybrids of rice (*Oryza sativa* L.) and barnyard grass (*Enchinochloa oryzicola* Vasing), *Mol Gen. Genet.,* 210, 39, 1987.
110. **Vasil, V. and Vasil, I. K.,** The ontogeny of somatic embryos of *Pennisetum americanum* (L.) K. Schum. I. In cultured immature embryos, *Bot. Gaz.,* 143, 454, 1982.
111. **Gray, D. J. and Conger, B. V.,** Somatic embryo ontogeny in tissue cultures of orchardgrass, in *Tissue Culture in Forestry and Agriculture,* Henke, R. R., Hughes, K. W., Constantin, M. J., and Hollaender, A., Eds., Plenum Press, New York, 49, 1985.
112. **Lu, C. Y. and Vasil, I. K.,** Histology of somatic embryogenesis in *Panicum maximum* (guinea grass), *Am. J. Bot.,* 72, 1908, 1985.
113. **Dunstan, D. I., Short, K. C., and Thomas, E.,** The anatomy of secondary morphogenesis in cultured scutellum tissues of *Sorghum bicolor, Protoplasma,* 97, 251, 1978.

114. **Haccius, B.,** Question of unicellular origin of non-zygotic embryos in callus cultures, *Phytomorphology,* 28, 74, 1978.
115. **Ozias-Akins, P. and Vasil, I. K.,** Proliferation of and plant regeneration from the epiblast of *Triticum aestivum* (wheat, Gramineae) embryos, *Am. J. Bot.,* 70, 1092, 1983.
116. **He, D. G., Tanner, G., and Scott, K. J.,** Somatic embryogenesis and morphogenesis in callus derived from the epiblast of immature embryos of wheat *(Triticum aestivum), Plant Sci.,* 45, 119, 1986.
117. **Gray, D. J. and Conger, B. V.,** Time-lapse light photomicrography and scanning electron microscopy of somatic embryo ontogeny from cultured leaves of *Dactylis glomerata* (Gramineae), *Trans. Am. Microsc. Soc.,* 104, 395, 1985.
118. **Brown, W. V.,** The grass embryo — a rebuttal, *Phytomorphology,* 15, 274, 1965.
119. **Ammirato, P. V.,** Organizational events during somatic embryogenesis, *Plant Tissue and Cell Culture,* Alan R. Liss, New York, 1987, 57.
120. **Jauhar, P. P.,** Genetic regulation of diploid-like chromosome pairing in the hexaploid species, *Festuca arundinacea* Schreb. and *F. ruba* L. (Gramineae), *Chromosoma,* 52, 363, 1975.
121. **Berg, C. C., Webster, G. T., and Jauhar, P. P.,** Cytogenetics and genetics, in *Tall Fescue,* Buckner, R. C. and Bush, L. P., Eds., American Society of Agronomy, Madison, WI, 1979, chap. 6.
122. **Buckner, R. C., Burris, P. B., Jr., and Todd, J. R.,** Morphological and chemical characteristics of inbred lines of tall fescue selected for palatability to cattle, *Crop Sci.,* 9, 581, 1969.
123. **Asay, K. H., Frakes, R. V., and Buckner, R. C.,** Breeding and cultivars, in *Tall Fescue,* Buckner, R. C. and Bush, L. P., Eds., American Society of Agronomy, Madison, WI, 1979, chap. 7.
124. **Larkin, P. J. and Scowcroft, W. R.,** Somaclonal variation. A novel source of variability from cell cultures for plant improvement, *Theor. Appl. Genet.,* 60, 197, 1981.
125. **Daub, M. E.,** Tissue culture and the selection of resistance to pathogens, *Ann. Rev. Phytopathol.,* 24, 159, 1986.
126. **Nabors, M. W., Gibbs, S. E., Bernstein, C. S.,** NaCl-tolerant tobacco plants from cultured cells, *Z. Pflanzenphysiol.,* 97, 13, 1980.
127. **Chaleff, R. S. and Keil, R. L.,** Genetic and physiological variability among cultured cells and regenerated plants of *Nicotiana tabacum, Mol. Gen. Genet.,* 181, 254, 1981.
128. **Kasperbauer, M. J. and Eizenga, G. C.,** Tall fescue doubled haploids via tissue culture and plant regeneration, *Crop Sci.,* 25, 1091, 1985.
129. **Brettell, R. I. S. and Thomas, E.,** Reversion of Texas male-sterile cytoplasm maize in culture to give fertile T-toxin resistant plants, *Theor. Appl. Genet.,* 58, 55, 1980.
130. **Gengenbach, B. G. and Rines, H. W.,** Use of phytotoxins in selection of disease resistant mutants in tissue culture, *Iowa State J. Res.,* 60, 449, 1986.
131. **Larkin, P. J. and Scowcroft, W. R.,** Somaclonal variation and eyespot toxin tolerance in sugarcane, *Plant Cell Tissue Organ Cult.,* 2, 111, 1983.
132. **Pauly, M. H., Shane, W. E., and Gengenbach, B. G.,** Selection for bacterial blight phytotoxin resistance in wheat tissue culture, *Crop Sci.,* 27, 340, 1987.
133. **Chapman, R. A.,** Diseases and nemtodes, in *Tall Fescue,* Buckner, R. C. and Bush, L. P., Eds., American Society of Agronomy, Madison, WI, 1979, chap. 15.
134. **Heinz, D. J., Krishnamurthi, M., Nickell, L. G., and Maretzki, A.,** Cell tissue and organ culture in sugarcane improvement, in *Applied and Fundamental Aspects of Plant Cell, Tissue and Organ Culture,* Reinert, J. and Bajaj, D. S., Eds., Springer-Verlag, Berlin, 1977, 3.
135. **Zehr, B. E., Williams, M. E., Duncan, D. R., and Widholm, J. M.,** Somaclonal variation in the progeny of plants regenerated from callus cultures of seven inbred lines of maize, *Can. J. Bot.,* 65, 491, 1987.
136. **Larkin, P. J., Ryan, S. A., Brettell, R. I. S., and Scowcroft, W. R.,** Heritable somaclonal variation in wheat, *Theor. Appl. Genet.,* 67, 443, 1984.

137. **Ahloowalia, B. S. and Sherington, J.,** Transmission of somaclonal variation in wheat, *Euphytica,* 34, 525, 1985.
138. **Ashraf, M., McNeilly, T., and Bradshaw, A. D.,** The response of selected salt-tolerant and normal lines of four grass species to NaCl in sand culture, *New Phytol.,* 104, 453, 1986.
139. **Sawhney, R. N.,** Induced mutagenesis for the development of multiline cultivars in bread wheat, *Euphytica,* 36, 49, 1987.
140. **Tabaeizadeh, A., Ferl, R. J., and Vasil, I. K.,** Somatic hybridization in the Gramineae: *Saccharum officinarum* L. (sugarcane) and *Pennisetum americanum* (L.) K. Schum. (pearl millet), *Proc. Natl. Acad. Sci. U.S.A.,* 88, 5616, 1986.
141. **Ozias-Akins, P., Ferl, R. J., and Vasil, I. K.,** Somatic hybridization in the Gramineae: *Pennisetum americanum* (L.) K. Schum. (pearl millet) + *Panicum maximum* Jacq. (guinea grass), *Mol. Gen. Genet.,* 203, 365, 1986.
142. **Fraley, R. T., Rogers, S. G., and Horsch, R. B.,** Genetic transformation in higher plants, *CRC Crit. Rev. Plant Sci.,* 4, 1, 1986.
143. **Grimsley, N., Hohn, T., Davies, J. W., and Hohn, B.,** *Agrobacterium*-mediated delivery of maize streak virus into maize plants, *Nature (London),* 325, 177, 1987.
144. **de la Pena, A., Lorz, H., and Schell, J.,** Transgenic rye plants obtained by injecting DNA into young floral tillers, *Nature (London),* 325, 274, 1987.
145. **Fuhr, G., Muller, T., Wagner, H., and Donath, E.,** Electroporation of oat protoplasts before and after fusion, *Plant Cell Physiol.,* 28, 549, 1987.
146. **Rhodes, C. A., Pierce, D. A., Mettler, I. J., Mascarenhas, D., and Detmer, J. J.,** Genetically transformed maize plants from protoplasts, *Science,* 240, 204, 1988.
147. **Vaeck, M., Reynaerts, A., Hofte, H., Jansen, S., De Beukeleer, M., Dean, C., Zabeau, M., Van Montagu, M., and Leemans, J.,** Transgenic plants protected from insect attack, *Nature (London),* 328, 33, 1987.
148. **Gray, D. J.,** Introduction to the sumposium, in Synthetic Seed Technology for the Mass Cloning of Crop Plants: Problems and Perspectives, *HortScience,* 22, 796, 1987.
149. **Fujii, J. A., Slade, D. T., Rendenbaugh, K., and Walker, K. A.,** Artificial seeds for plant propagation, *Trends Biotechnol.,* 5, 335, 1987.
150. **Stuart, D. A., Strickland, S. G., and Walker, K. A.,** Bioreactor production of alfalfa somatic embryos, in Synthetic Seed Technology for the Mass Cloning of Crop Plants: Problems and Perspectives, *HortScience,* 22, 803, 1987.
151. **Redenbaugh, K., Slade, D., Viss, P., and Fujii, J. A.,** Encapsulation of somatic embryos in synthetic seed coats, in Synthetic Seed Technology for the Mass Cloning of Crop Plants: Problems and Perspectives, *HortScience,* 22, 803, 1987.
152. **Gray, D. J.,** Quiescence in monocotyledonous and dicotyledonous somatic embryos induced by dehydration, in Synthetic Seed Technology for the Mass Cloning of Crop Plants: Problems and Perspectives, *HortScience,* 22, 810, 1987.
153. **Gray, D. J., Conger, B. V., and Songstad, D. E.,** Desiccated quiescent somatic embryos of orchardgrass for use as synthetic seeds, *In Vitro Cell. Dev. Biol.,* 23, 29, 1987.

Chapter 4

PLANT REGENERATION AND EVALUATION

M. J. Kasperbauer

TABLE OF CONTENTS

I. INTRODUCTION

Genetic improvement of tall fescue (*Festuca arundinacea*) by conventional plant breeding is very slow because the species is wind-pollinated and generally self-sterile. Because of these factors, it is not possible to rapidly develop inbreds to concentrate desired genes for use in development of new hybrids or cultivars. Therefore, development and use of unconventional methods utilizing cell and tissue culture offer a possible approach to speed up improvement of tall fescue and its hybrids. This chapter will concentrate on development and evaluation of plant regeneration procedures, and on evaluation of somatic tissue-derived plants. Much of the discussion is based on the research of the author and Dr. R. C. Buckner that began in the mid 1970s.

Working models were developed for potential use of cell and tissue culture to speed improvement of tall fescue and an intergeneric annual ryegrass (*Lolium multiflorum*) × tall fescue hybrid (Figure 1). The first callus cultures were established in 1975. Plant regeneration procedures were soon developed by applying principles already known to function in initiation of shoots from tobacco (*Nicotiana tabacum*) callus. That is, tobacco callus will initiate leafy shoots if the auxin/cytokinin ratio is decreased in the culture medium. Since the tall fescue and the hybrid callus cultures did not require cytokinin in the medium, we reasoned that fescue callus was autonomous for cytokinin. If true, we hypothesized that reduction of the auxin (2,4-D) concentration should result in shoot regeneration. It did. Field plot evaluations of somatic tissue-derived plants began in April, 1976, in Dr. Buckner's fescue nursery near Lexington, KY. The experimental approach was subsequently expanded to development and potential use of haploids and doubled haploids in tall fescue improvement, as discussed in Chapters 5 and 6 of this book.

II. PLANT REGENERATION

The ability to regenerate plants from genetically altered cells is critical in development of potentially useful biotechnological approaches to plant genetic improvement. Also, regeneration of numerous genetically identical plants (clones) from potentially superior plants derived from cell cultures (or from conventionally derived plants) is an important part of the fescue germplasm evaluation process. For complete evaluation of potentially useful genotypes, it would be desirable to have enough genetically identical plants to allow simultaneous evaluation for a number of characteristics such as chromosome pairing, forage (and turf) productivity and quality, disease resistance, and tolerance to cold, drought, or heat stress.

Tillering is the natural form of vegetative propagation (cloning) in tall fescue. Therefore, a basic understanding of the tillering process and its regulation should provide some insight for development of methods for plant regeneration from cell and tissue cultures.

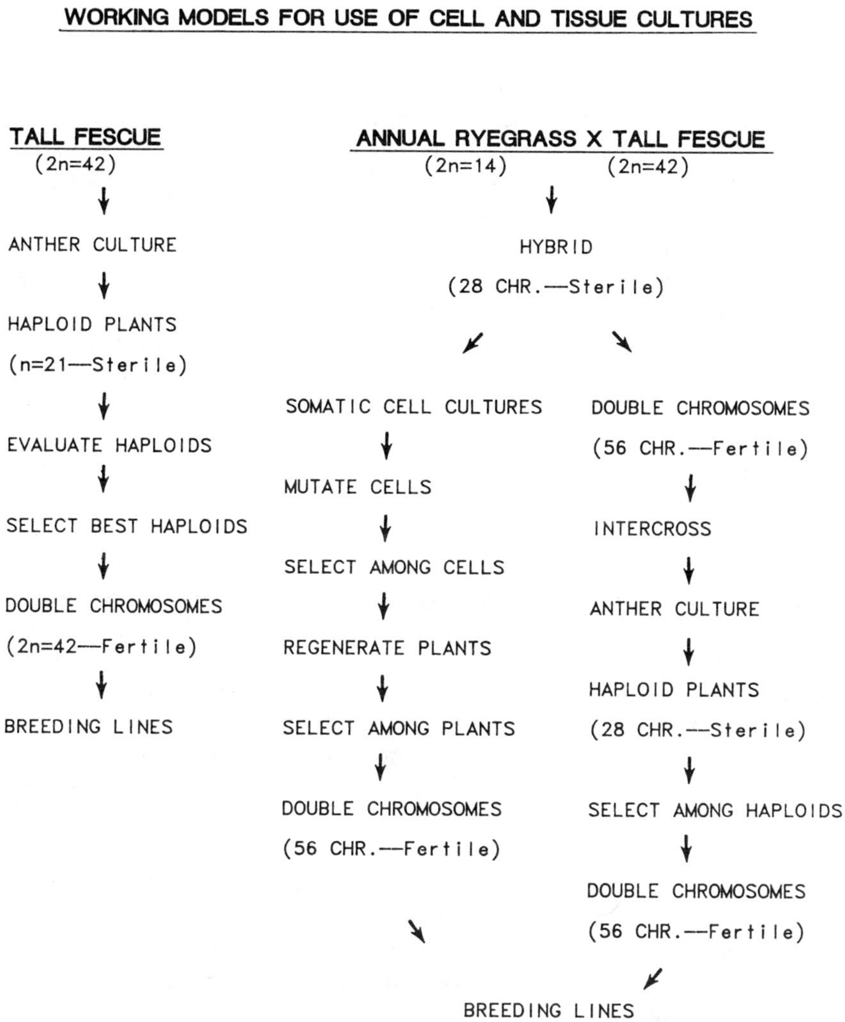

FIGURE 1. Working model for use of cell and tissue culture to speed up improvement of tall fescue and annual ryegrass × tall fescue hybrids. (Adapted from Kasperbauer, M. J., Buckner, R. C., and Bush, L. P., *Crop. Sci.*, 19, 458, 1979.)

A. Tillering

In nature, tall fescue flowers during the long days of spring and early summer, sets seed, goes through "summer dormancy", and then develops tillers during the shortening days of autumn. It is important to realize that tillering in a grass plant is analogous to branching in a broadleaf plant such as soybean (*Glycine max*). That is, tillers develop from axillary buds. Thus, the probability is extremely high that all tillers from the same plant are genetically identical.

The amount of tillering is influenced by the genetics of the plant and the environment under which it is growing. Certain environmental factors are sensed by the plant and various genes are activated or repressed to modify plant development to favor plant survival in the sensed environment.[1] Day length (photoperiod) and the population density of competing plants are major contributors to the environmental aspects. Nutritional status of the plant also contributes to the abundance of tillers, or continued development of tillers after they are initiated.[2]

A tall fescue plant is able to sense decreasing day lengths and initiate vegetative reproduction of itself via tillers. The plant is also able to sense the number and nearness of competing plants, and to autoregulate the amount of tillering. For example, an isolated plant will develop more tillers than will a genetically identical one growing among many other plants, even if all other environmental factors remain constant.

The ability of a plant to sense nearness of competing plants, even before mutual shading occurs, is dependent upon the light absorption and reflection characteristics of leaves. Each green leaf absorbs most of the visible light and reflects or transmits most of the far-red light that it receives. Therefore, the number and nearness of competing leaves influences the amount of reflected far-red light and the ratio of far-red relative to red light received by a plant.[3] The photoequilibrium status of the phytochrome system within the plant is influenced by the far-red/red light ratio.[4] The phytochrome system regulates many developmental processes,[5] such as tillering.[6] A low far-red/red light ratio results in increased tillering, and vice versa. The amount of reflected far-red and the far-red/red light ratio are low when seedlings are small and/or far apart. Under such conditions, the relatively low far-red/red ratio (acting through the phytochrome system) promotes initiation of new tillers. Then, as the tillers increase in number and size, they reflect more far-red light and the far-red/red ratio increases to the point that it stops further tillering. This natural regulatory mechanism warrants further discussion as a useful procedure and as a background in development of *in vitro* cell culture and plant regeneration procedures.

Limited controlled environment study of the environmental influence on tillering of tall fescue has shown that tillering is enhanced under moderately cool, 12 h (relative to longer) photoperiods of bright light that end with a low ratio of far-red relative to red light.[7] Although in-depth controlled-environment studies of tillering have not been published for tall fescue and its hybrids, very relevant studies of tillering have been reported for winter wheat (*Triticum aestivum*).[6] Since results of preliminary unpublished studies with tall fescue were very similar to those conducted with winter wheat, results obtained with wheat are presented in this chapter.

As shown in Table 1, the number of tillers per plant was much higher on wide-spaced field-grown seedlings. Also, the number of tillers per plant at maturity was less than in the seedling stage within each environment. These results are consistent with the concepts discussed above.

TABLE 1
Tillers per Wheat Plant Grown in Wide, Intermediate, and Close (30, 120, and 300 Seedlings per Square Meter) Plantings in a Field

	Plant spacing		
Plant stage	**Wide**	**Intermediate**	**Close**
Seedling	14.0	—	2.9
Mature	9.8	3.5	1.9

Adapted from Kasperbauer, M. J. and Karlen, D. L., *Physiol. Plant.*, 66, 162, 1986.

TABLE 2
Photon Ratios of Far-Red Relative to Red Light Received by Wheat Seedlings Grown in Wide, Intermediate, and Close-Spaced Field Populations (Measurements Taken in Late Afternoon)

Direct sunlight (control)	Plant spacing		
	Wide	**Intermediate**	**Close**
0.91	1.15	1.44	1.72

From Kasperbauer, M. J. and Karlen, D. L., *Physiol. Plant.*, 66, 161, 1986.

The photon ratios of far-red/red light received by seedlings in different population densities are shown in Table 2. The ratio was higher in close- than in wide-spaced plants. These readings were all made in sun flecks (i.e., not in shade). Therefore, it is of significance that even the ratio in the wide-spaced population is appreciably higher than that in direct sunlight, away from the plants. It is relevant that earlier studies of plant developmental effects of various wavelengths of light showed that seedlings are very sensitive to even subtle changes in the far-red/red light ratio.[8]

A controlled-environment experiment was conducted with winter wheat seedlings starting a few days after emergence. They received brief exposure to red or far-red at the end of the daily photosynthetic period, and this clearly influenced plant development. Exposure to red (a low far-red/red ratio) resulted in development of more tillers, shorter leaves, and a larger root system than developed on similar plants that received far-red (a high far-red/red ratio) at the end of the daily photosynthetic period (Table 3). These results were consistent with the field responses (Tables 1 and 2). The response to a brief

TABLE 3
Effects of Phytochrome on Tillering, Leaf Length, and Shoot/Root Biomass Ratios of Wheat Seedlings in Controlled Environment

Plant character	Light at end of 12-h day		
	R	FR	FR + R
Tillers/plant (no.)	8.7 ± 0.3	5.4 ± 0.3	8.1 ± 0.2
Leaf length (mm)	425 ± 5	460 ± 5	420 ± 5
Shoot/root biomass (ratio)	1.09	1.36	1.10

Note: Red (R) and far-red (FR) light treatments were for 5 min at the end of daily 12-h photosynthetic periods for 20 consecutive days. The FR + R treatment consisted of 5 min of FR followed immediately by 5 min of R to test photoreversibility of the control system.

From Kasperbauer, M. J. and Karlen, D. L., *Physiol. Plant.*, 66, 162, 1986.

exposure to red immediately following far-red is evidence of photoreversibility of the light receptor and demonstrates that the receptor is phytochrome.

It is clearly evident that tillering is regulated by environmental sensors within the plant that measure day length and spectral distribution of light, and then regulate developmental responses to favor survival of the plant. That is, tillering would lead to production of more potential flowering stems which would be favorable up to a point; however, the plant can stop further tillering when the sensing system determines that continued tiller formation would be counterproductive.

The important points in this discussion are that day length (photoperiod) signals the plant to produce tillers to clone itself, and the far-red/red ratio (as is influenced by the amount of reflected far-red light as a measure of size, number, and nearness of competing plants) acts through the phytochrome system in autoregulation of amount of tillering. It is apparent that the phytochrome system functions as a variable switch that constantly monitors the far-red/red ratio (as a measure of competition) and regulates tillering as well as other developmental processes.

Since natural tillering results in vegetative propagation of a relatively few (although genetically identical) plants, it would be reasonable to draw on this information to develop cell and/or tissue culture methods that could result in large numbers of genetically identical plants from a superior field-grown plant. Such methods would be useful whether the plant was derived through conventional plant breeding or through biotechnological approaches.

B. Plant Regeneration from Cell and Tissue Cultures

A number of experiments designed to regenerate plants from somatic cells were conducted with tall fescue and an intergeneric annual ryegrass × tall fescue hybrid that was developed by Dr. Buckner. The hybrid was an attempt to combine forage qualities of ryegrass and the persistence of fescue.[9] The F_1 hybrid has 7 chromosomes from annual ryegrass and 21 chromosomes from tall fescue. The 28-chromosome hybrid is not fertile. Chromosome doubling is necessary to obtain fertility. Cell and tissue culture approaches were introduced as part of the research and development program.

Since many of the pioneering studies of cell and tissue culture were done with tobacco, experience with that species served as a basis for development of tissue preparation methodology, media formulation, and physical environments selected for initial studies with tall fescue and its hybrid. In addition to the development of basic nutrient media for cell cultures,[10,11] some of the early studies demonstrated that tobacco callus tissue could be induced to form shoots or roots by changing the auxin/cytokinin ratio in the nutrient medium. Other early research utilized tobacco tissue cultures to demonstrate that undifferentiated callus contained phytochrome and that the ratio of far-red/red light received by the callus acted through the phytochrome system within the cells to influence metabolic processes.[12,13] Callus that received red light for 5 min (alternated with 23 h and 55 min of darkness) per day accumulated significantly more dry matter during a 14-d treatment period.[12] The far-red/red photoreversible response in undifferentiated callus cultures was reported in 1966.[12] It was the first evidence that phytochrome is present in and can regulate metabolic processes in undifferentiated callus on artificial nutrient medium. The fact that phytochrome is present in dividing and recently divided cultured cells is relevant in selecting a light source, or darkness, for use during cell and tissue cultures and during plant regeneration from those cultures. Photodestructability of various components of the medium should also be considered when selecting a light regime for tissue culture and for plant regeneration from cell and/or tissue cultures. Some of these apparently subtle differences brought about by exposure to light may be major factors in success or failure in plant regeneration from cell and tissue cultures. After considering the foregoing factors, a number of experiments were conducted with tall fescue and the intergeneric hybrid to develop cell and tissue culture and plant regeneration methods analogous to those already proven successful for tobacco.

1. Explant Source

a. Embryos

Immature embryos and/or embryos from germinating seeds are among the most convenient materials from which to establish cultures. Embryos are relatively easy to surface decontaminate, establish callus, and regenerate plants. Embryo-derived cultures are useful for development of medium requirements and physical environment conditions for a given species. The plants regenerated from embryo-derived tissue cultures have a high probability (relative to leaf- or stem-derived callus) of being genetic clones because of the nature of the explant source. However, a major disadvantage of cloning a highly heterozygous species, such as tall fescue, from embryo cultures is that there is a very high probability that each embryo is genetically different from all other embryos. Therefore, there is no assurance that a randomly selected embryo would contain desired basic forage and/or turf characteristics and cold, heat, or drought tolerance. To obtain that and other relevant information, it would be desirable to evaluate the plants under field conditions to eliminate undesirable genotypes before establishment of cell and/or tissue cultures. For example, incorporation of improved forage (or turf) characteristics via cell or tissue culture would be much more useful if incorporated into a plant that is field hardy. Therefore, establishment of aseptic tissue cultures from field-grown plants is a desirable approach. However, it is much more difficult than establishment from embryos.

b. Field-Grown Plants

A number of experiments were conducted with explants from field-grown plants to compare explant source, medium additives, physical environment, and timing of subcultures. Critical factors and rationale for various approaches are discussed below.

Field-grown plants allow preselection of source plants for growth, development, and survival characteristics before establishing the cell cultures. Of course, a field-grown plant has a much greater probability of microbial contamination on the leaf or stem surface as well as within the tissue. This fact increases the importance of surface decontamination of the explants. It is important to kill the microbial contaminants without killing the critical plant cells. A slight modification of procedures used for field-grown tobacco[14,15] was successful for field-grown tall fescue and its hybrids.

Vigorously growing plants were selected during rapid spring growth when panicles began to emerge. Leaf, stem, and peduncle explants were selected and cut into 2- to 5-cm segments. These were surface decontaminated by washing for about 6 min in 15% laundry bleach and 85% water (i.e., about 0.8% sodium hypochlorite in water). All subsequent steps were done in a sterile transfer hood. The decontamination process was followed by three brief rinses in sterile water to remove the remaining sodium hypochlorite from the tissues. The selected somatic explants were then cut to 2- to 5-mm segments, trimmed of extraneous tissue, and placed on appropriate culture medium. The explants were cut and trimmed under sterile water to avoid drying at the cut surfaces before placement on the medium.

2. Culture Media

Although a number of medium formulations have been used for cell and tissue cultures of various species, we used the same basic medium for tall fescue and its hybrids as used for tobacco tissue cultures. The medium formulation, prepared as six stock solutions (Table 4), is based on our modification[13,15] of Murashige and Skoog[10] and Linsmaier and Skoog.[11] The

TABLE 4
Composition of Basic Fescue Culture Media

Constituent	Stock solution[a] Code	Stock solution[a] g/l	Culture medium[b] ml/stock l medium	Culture medium[b] mg constituent/ l medium
Medium I. Somatic Callus Establishment and Maintenance, and First-Stage of Anther-Panicle Culture				
EDTA	A	0.80	40	32.0
$Fe_2(SO_4)_3$		0.38		15.2
NH_4NO_3	B	82.50	20	1650
KNO_3		95.00		1900
H_3BO_3	C	1.24	5	6.2
KH_2PO_4		34.00		170.0
KI		0.166		0.83
$Na_2MoO_4 \cdot 2H_2O$		0.050		0.25
$CoCl_2 \cdot 6H_2O$		0.005		0.025
$MgSO_4 \cdot 7H_2O$	D	74.00	5	370
$MnSO_4 \cdot H_2O$		4.46		22.3
$ZnSO_4 \cdot 7H_2O$		1.72		8.6
$CuSO_4 \cdot 5H_2O$		0.005		0.025
$CaCl_2 \cdot 2H_2O$	E	88.00	5	440
Thiamine · HCl	F	0.02	5	0.1
Nicotinic acid		0.10		0.5
Pyridoxine · HCl		0.10		0.5
Glycine		0.40		2.0
2,4-D	—	0.20	10	2.0
Sucrose	—	—	—	20,000
Agar	—	—	—	6,000
myo-Inositol	—	—	—	100

Medium II. Plant Regeneration from Callus, and Second-Stage of Anther-Panicle Culture

Composition is identical with Medium I, except that 1.25 ml of 2,4-D stock is used giving a final concentration of 0.25 mg/l

Medium III. Rooting

Composition is identical with Medium I, except that no 2,4-D is used, agar is reduced to 5,000 mg/l, and all other components are half strength

[a] Bring components to volume with glass distilled water. Because of the small amount of 2,4-D used, a smaller amount of stock solution is prepared. The 2,4-D is first dissolved in a few drops of ethyl alcohol, then added slowly to the glass-distilled water.

TABLE 4 (continued)
Composition of Basic Fescue Culture Media

[b] Add stock solutions and all weighed components (sucrose, agar and inositol) to glass-distilled water, then bring to final volume with glass-distilled water. Heat on stir-hotplate until agar melts. Adjust to pH 5.7 with 2 *N* NaOH. Autoclave at 121°C for 15 min. Cool and pour culture vessels before agar solidifies.

Adapted from Kasperbauer, M. J. and Wilson, H. M., in *Nicotiana: Procedures for Experimental Use* (USDA Tech. Bull. 1586), Durbin, R. D., Ed., U.S. Department of Agriculture, Washington, DC, 1979, chap. 5.

present author considered this to be a logical approach because whole plants of both species have the same basic nutrient requirements, and the objective was to accomplish with tall fescue and its hybrids what was already accomplished with tobacco (i.e., establish and maintain callus cultures, culture haploids, and regenerate doubled haploids from somatic tissue excised from haploid plants).[15] The only significant difference between media used for fescue and tobacco was the amount of auxin and cytokinin in the medium.

3. Tissue Source

Explants from leaf veins, elongated stems, and soft tissue from the base of peduncles and internodes were cultured on the basic medium supplemented with a range of auxin and cytokinin concentrations. As shown in Table 5, no callus formed on leaf and stem segments that had already differentiated on the source plant. This was in contrast to tobacco in which cell cultures were easily established from fully expanded leaves and from stem pith.[15] Many of the explants from fully expanded fescue stem tissue from field-grown plants also contained a slow-growing (in culture) endophyte identified as *Epichloe typhina* (now known as *Acremonium coenophialum*). An important finding was that callus formed only from the very young tissue at the base of rapidly elongating peduncles and internodes when cultured on medium containing 2 to 20 mg of 2,4-D per liter of medium. A concentration of 2 mg of 2,4-D per liter was subsequently used as the standard level for callus establishment and maintenance because this level was effective and because higher levels appeared to have undesirable side effects. An added bonus was that the callus established from the youngest tissue was not contaminated with the endophyte. This is analogous with earlier research that resulted in disease-free plants through meristem culture.[16] The results shown in Table 5 provided several important items of information: (1) the medium formulation allowed callus establishment from explants that had many dividing or recently divided cells, (2) callus did not form from expanded leaf or stem explants (this became a very important factor for culture of tall fescue haploids using panicle tissue as a nurse culture that was not capable of callus formation or plant regen-

TABLE 5
Callus Formation on Stem and Leaf Explants of Annual Ryegrass × Tall Fescue F_1 Hybrids Cultured on Initiation Medium Supplemented with Various Levels of 2,4-D, NAA, and Kinetin

Medium supplement (mg/l)			Explant source and callus formation[a]		
				Internode tissue	
2,4-D	NAA	Kinetin	Leaf vein	Elongating	Elongated
0.0	0.0	0.0	–	–	–
0.0	0.1	0.2	–	–	–
0.0	0.1	2.0	–	–	–
0.0	2.0	0.2	–	–	–
0.2	0.0	0.0	–	–	–
0.5	0.0	0.0	–	±	–
0.5	0.0	0.2	–	+	–
1.0	0.0	0.0	–	+ +	–
1.0	0.0	0.2	–	+ +	–
2.0	0.0	0.0	–	+ + +	–
2.0	0.0	0.2	–	+ +	–
4.0	0.0	0.0	–	+ +	–
4.0	0.0	0.2	–	+ +	–

[a] – indicates no callus formation and ±, +, + +, and + + + indicate callus formation and relative amount.

From Kasperbauer, M. J., Buckner, R. C., and Bush, L. P., *Crop Sci.*, 19, 458, 1979.

eration, as discussed in Chapter 5 of this book), and (3) the auxin requirement was much higher, and the cytokinin requirement was lower for fescue than for tobacco. In fact, cytokinin was not needed in the medium for callus establishment, maintenance, or plant regeneration.

4. Physical Environment

Because tissue culture studies with tobacco had shown that the cells contain phytochrome and exhibit a growth response to the ratio of far-red relative to red light,[12,13] fescue explants from young peduncle and stem tissue were cultured under various lighting regimes. As shown in Table 6, there was no dramatic difference in percentages of explants that formed callus in light vs. darkness. Although some differences were apparent among genetic lines, all formed some callus. Awareness that callusing occurred in both light and darkness allowed flexibility in development of protocols, such as maintenance of cultures in a dark cabinet (with brief daily illuminations during inspections). In subsequent experiments, it was determined that low-intensity

TABLE 6
Callus Formation on Explants from Elongating Internodes and Peduncles of Annual Ryegrass × Tall Fescue F_1 Hybrids and Parental Lines at 25 ± 1°C in Light or Darkness on Media Containing 2 or 4 mg of 2,4-D per Liter

	Callus formation on explants			
	Light		Darkness	
Source of explants	No. of calli/ no. of explants	%	No. of calli/ no. of explants	%
Parental lines				
Annual ryegrass	—	—	7/12	58
Tall fescue	—	—	8/13	62
F_1 hybrid lines				
13—1	14/18	78	12/19	63
20—1	9/21	43	16/27	59
20—5	30/49	61	20/29	69
38—9	7/32	22	5/24	21
38—15	13/15	87	19/22	86
38—19	13/25	52	12/21	57
38—20	8/22	36	18/23	78
58—12	9/17	53	10/18	56
58—30	22/37	59	18/30	60
58—32	18/22	82	15/20	75

From Kasperbauer, M. J., Buckner, R. C., And Bush, L. P., *Crop Sci.*, 19, 458, 1979.

light from cool white fluorescent lamps was useful during plant regeneration from callus because it kept the regenerating plants from becoming etiolated.

5. Plant Regeneration

Numerous plants were regenerated from tall fescue and from annual ryegrass × tall fescue hybrid callus tissue by decreasing the 2,4-D concentration of the medium to 0.00 or 0.25 mg/l (Table 7). Preliminary experiments using the basic medium with the decreased 2,4-D level showed that plant regeneration occurred in darkness or in light within 3 to 4 weeks. Plants were routinely regenerated from callus under low-intensity light from cool white fluorescent lamps in order to avoid etiolation. Cool white fluorescent lamps were used because they emit a low ratio of far-red relative to red light, which acts through the phytochrome system to suppress shoot elongation and favor rooting (see Tables 1 to 3).

Complete removal of 2,4-D from the regeneration medium resulted in no new callus formation and many regenerated plants with abundant roots. This combination resulted in tangled roots and much difficulty in separating the plants. Consequently, many plants were damaged during separation and transfer to starting medium. On the other hand, the 0.25 mg level of 2,4-D resulted

TABLE 7
Callus and Plant Regeneration Response of Tall Fescue Callus Cultures to 2,4-D Level after Establishment and Maintenance on Basic Medium with 2 mg 2,4-D per Liter

	Response[a]	
2,4-D conc.	**Callus**	**Plantlets**
(mg/l)		
4.0	+ + +	−
2.0	+ + +	±
0.25	+	+ + +
0.0	−	+ + +

[a] −, ±, +, and + + + indicate relative amounts

Adapted from Kasperbauer, M. J., Buckner, R. C., and Bush, L. P., *Crop Sci.*, 19, 458, 1979.

in formation of some new callus and regeneration of many plants, with very few roots. These plants were easy to separate when they were about 1 cm tall. Very few were damaged, and they rooted quickly when placed on a rooting medium (half-strength basic medium [Table 4] with 5 g of agar per liter and no 2,4-D). Rooting and plant establishment were done in a 21 ± 1°C growth chamber with light from cool white fluorescent lamps. When the roots were about 1 cm long, individual plants were coded and root tips were collected and processed for cytological evaluation (see Chapter 7 of this book). The plants were returned to sterile rooting medium until new roots were about 1 to 2 cm long, after which the plants were transferred to sterile potting soil. The newly potted plants were covered with transparent plastic to maintain high humidity for 3 d. The starting chamber was controlled at 21 ± 1°C and illuminated for 12 h/d with cool white fluorescent light at about 200 μmol m^{-2} s^{-1}. After the 3rd d, the plastic cover was removed for a few hours each day to condition the plants to lower humidity. The plastic cover was completely removed after another 3 d. The light level was gradually brought up to that of full sunlight, and plants were transferred to a greenhouse. After establishment of the regenerated plants under greenhouse conditions, more than 1200 of them were taken to field plots for evaluation in conjunction with Dr. Buckner's tall fescue improvement program.

Tillers were taken from some of the field-tested plants and grown in a controlled environment selected to produce numerous new tillers (i.e., 12-h days from cool white fluorescent lamps [low far-red/red ratio] at 21 ± 1°C). The tillers were used as a source of somatic tissue for cytological evaluation.

III. EVALUATION OF REGENERATED PLANTS

Successful regeneration of plants from cell and tissue cultures is an important step. However, thorough evaluation is needed before they are usable in a plant improvement program.

A. Visual

Early visual evaluation of newly regenerated shoots can identify obvious albinos, which occur in greater frequency when the regeneration medium contains a high concentration of sucrose.[17,18] Other abnormalities, such as highly abnormal leaves and failure to develop roots, are also identifiable during the rooting phase, before plants are transplanted to potting soil.

B. Cytological

Cytological evaluations (discussed in detail by Springer in Chapter 7 of this book) should involve determination of chromosome numbers, pairing, and other possible abnormalities. Determination of chromosome numbers is an important step in determining potential usefulness of the regenerated plants. After the initial screening for chromosome number and evaluation of growth characteristics, more detailed analyses of chromosome pairing should be done on the plants that are considered to be the best candidates for incorporation into the breeding program (see Chapter 7 for further discussion).

C. Field

After preliminary visual and cytological screening to reduce the numbers, the regenerated plants should be grown under field conditions to evaluate summer heat tolerance, winter cold hardiness, and growth characteristics that may determine usefulness of the genotype. Forage and/or turf characteristics should be determined for each candidate plant. The author's experience with field evaluation of regenerated plants is discussed below.

In order to have sufficient replicated samples for the field studies, we first grew the regenerated plants in pots under a controlled environment that caused profuse tillering. The vegetatively propagated plants were then coded and planted in replicated field plots for evaluation.

The first group of about 1200 regenerated plants were started in pots in a greenhouse and then moved to field plots in April 1976. They were evaluated over a 2-year period. Field evaluations during subsequent years were done with smaller numbers of preselected plants and for various durations, depending on the objectives.

The plants shown in Figure 2 were photographed in late spring, following their first winter in field plots. Nearly all plants survived the winter, and many of them appeared to be identical. A key point became apparent during the evaluation. The number and duration of subcultures prior to plant regeneration was important. Plants that were regenerated from callus after two 3-

FIGURE 2. Annual ryegrass × tall fescue plants that were regenerated from callus cultures.

to 4-week subcultures on the maintenance medium (2 mg of 2,4-D per liter) were compared with plants regenerated after five 3- to 4-week subcultures and with others regenerated after one or two 3- to 4-week subcultures and one 8-week subculture before plant regeneration. Plant regeneration proficiency and the general characteristics of the regenerated plants obtained by these three procedures are discussed below. The general pattern of results provided the basis for subsequent research to develop procedures for successful culture of tall fescue haploids (Chapter 5 of this book) and regeneration of doubled haploids from cytologically verified haploids (Chapter 6 of this book).

1. Plants Regenerated after Two Brief Subcultures

Many normal-appearing shoots developed and appeared to be clonal reproductions of the original plants. Root tip cytology showed that all of the regenerated plants also had the same chromosome number as the original plants. Thus, under the conditions used, callus could be established from young stem tissue from tall fescue and the annual ryegrass × tall fescue hybrid, and normal plants could be regenerated from the callus after two brief subculture periods on maintenance medium. This procedure showed promise for rapid clonal increase of a plant that exhibits outstanding characteristics in

FIGURE 3. Normal-, narrow-, and wide-leaved plants regenerated from the fifth subculture of callus established from the same young stem explant from annual ryegrass × tall fescue F_1 hybrid line 58-12. (From Kasperbauer, M. J., Buckner, R. C., and Bush, L. P., *Crop Sci.*, 19, 458, 1979.)

a field evaluation. Such rapidly produced clones developed from selected plants could be very useful to increase plant numbers for use in detailed physiological studies in which the same genotype must be evaluated under a number of different environments, or be sampled at different developmental stages.

2. *Plants Regenerated after Five Brief Subcultures*

Calli that went through five successive 3- to 4-week subcultures before being placed on the plant regeneration medium developed fewer shoots per mass of callus than those that went through only two 3- to 4-week subcultures before plant regeneration. Also, a higher percentage of the plants regenerated from the ''five-subculture'' series were phenotypically variant from the original source plant. In one study with tissue excised from an annual ryegrass × tall fescue F_1 hybrid plant, a small percentage of the regenerated plants had very fine leaves, and an equally small percentage had wider than normal leaves (see Figure 3) even though plants of all three phenotypes had the chromosome number of the source plant. These variants were identified in 1975 and have been vegetatively propagated through tillers until the present time (1989) for possible further use in some aspect of forage or turf improvement. The plants have been evaluated for forage characteristics and for summer and winter hardiness in Kentucky and South Carolina.

Subsequent experiments with tall fescue also showed that as the number of subcultures increased, the ability to regenerate plants decreased and the percentage of phenotypic variants (also called somaclonal variants) increased. Recent examination of a number of regenerated tall fescue plants for root characteristics showed that some of the regenerated lines had distinctly different root systems, even though shoots appeared the same and chromosome numbers did not differ. We are now (1989) researching the possible use of cell and tissue culture to develop plants with superior root systems. For example, roots with the ability to penetrate a high bulk-density subsoil layer would be a major advantage on soils with low natural fertility and low water-holding capacity in the surface layer. These experiments have shown that the percentage of phenotypic variants could be increased by increasing the number of subcultures before regenerating plants. Conversely, a large number of apparently identical plants could be obtained by regenerating plants after only one or two brief subcultures.

3. Plants Regenerated after One Brief and One Extended Subculture

This approach was a preliminary attempt to mimic some tobacco culture responses and regenerate plants with higher ploidy than the source plant. With tobacco, it is possible to excise midvein tissue from a fully expanded leaf and to regenerate plants with the doubled chromosome number directly from the leaf tissue. A modified approach was taken with fescue because earlier attempts to culture plants, or to establish callus, from fully expanded ryegrass × fescue leaf or stem tissue were unsuccessful (Table 5). We reasoned that a possible alternative with fescue was to establish callus from young stem tissue (Table 5) and, after increasing the mass of callus during one or two subcultures, to "age" the callus during an extended subculture before regenerating plants. We hypothesized that the same chain of events that took place to induce endomitotic divisions followed by regeneration of plants from "aged" tissue taken from fully expanded tobacco leaves[14,15] might be duplicated by "aging" the fescue callus on maintenance medium before regenerating plants. The first experiments were done with callus derived from the annual ryegrass × tall fescue F_1 hybrid. These plants had 28 chromosomes (7 from the ryegrass and 21 from the fescue parent). The F_1 hybrid (like a tobacco haploid) was infertile, and plants with the doubled number of chromosomes might be fertile (like a tobacco doubled haploid). The initial attempts with "aged" callus from the F_1 hybrid resulted in 111 regenerated plants that were grown as individual entries in field evaluation plots. Cytological analyses were not completed before the field evaluation because of a scheduling problem. All plants survived the first winter. Although most of the regenerated plants had the same general phenotypic characteristics as the F_1 source plant,

11 of them displayed visible amphiploid characteristics[9] such as larger stems and wider leaves. Root tip analyses of the 11 putative amphiploids showed that 9 of them had the doubled number of chromosomes, 1 had the F_1 number, and the other one was not countable. This preliminary experiment demonstrated that endomitotic divisions could be stimulated in such callus and that plants with the increased chromosome number could be regenerated. This information provided the basis for later experiments in which tall fescue doubled haploids were cultured from callus derived from haploids (Chapter 6 of this book). The fact that not all of the regenerated plants with visually detected amphiploid characteristics had the doubled chromosome number emphasizes the importance of using cytological procedures as part of the overall evaluation of regenerated plants that are derived through various biotechnological methods.

IV. SUMMARY

Observation of tillering responses of fescue to various environmental conditions, and experience with tobacco tissue culture and plant regeneration, provided model systems for development of tall fescue tissue culture and plant regeneration. Although callus cultures are more easily established from immature embryos than from field-grown fescue plants, field-grown plants are prescreened for a number of characteristics such as winter or summer hardiness. Thus, this chapter concentrated on establishment and maintenance of callus cultures from field-grown fescue, plant regeneration, and various facets of evaluation. It also attempted to show how successful use of fescue cell and tissue cultures can contribute to more rapid isolation of pure lines that can speed up genetic improvement of this important wind-pollinated, self-infertile, polyploid perennial grass species.

REFERENCES

1. **Kasperbauer, M. J.,** Phytochrome involvement in regulation of the photosynthetic apparatus and plant adaptation, *Plant Physiol. Biochem.*, 26(4), 519, 1988.
2. **Simmons, S. R. and Lauer, J. G.,** Shoot photoassimilate partitioning patterns during the tillering phase in spring barley, in *Phloem Transport,* Cronshaw, J., Lucas, W. J., and Giaquinta, R. T., Eds., Alan R. Liss, New York, 1986, 519.
3. **Kasperbauer, M. J.,** Far-red reflection from green leaves and effects on phytochrome-mediated assimilate partitioning under field conditions, *Plant Physiol.*, 85, 350, 1987.
4. **Hendricks, S. B., Butler, W. L., and Siegelman, H. W.,** A reversible photoreaction regulating plant growth, *J. Phys. Chem.*, 66, 2550, 1962.
5. **Borthwick, H. A.,** Phytochrome action and its time displays, *Am. Nat.*, 98, 347, 1964.
6. **Kasperbauer, M. J. and Karlen, D. L.,** Light-mediated bioregulation of tillering and photosynthate partitioning in wheat, *Physiol. Plant.*, 66, 159, 1986.

7. **Kasperbauer, M. J., Buckner, R. C., and Bush, L. P.,** Tissue culture of annual ryegrass × tall fescue F_1 hybrids: Callus establishment and plant regeneration, *Crop Sci.*, 19, 457, 1979.
8. **Kasperbauer, M. J., Borthwick, H. A., and Hendricks, S. B.,** Reversion of phytochrome 730 (Pfr) to P660 (Pr) in *Chenopodium rubrum* L., *Bot. Gaz.*, 125(2), 75, 1964.
9. **Buckner, R. C., Hill, H. D., and Burrus, P. B.,** Some characteristics of perennial and annual ryegrass × tall fescue hybrids and of the amphidiploid progenies of annual ryegrass × tall fescue, *Crop Sci.*, 1, 75, 1961.
10. **Murashige, T. and Skoog, F.,** A revised medium for rapid growth and bioassays with tobacco tissue cultures, *Physiol. Plant.*, 15, 473, 1962.
11. **Linsmaier, E. M. and Skoog, F.,** Organic growth factor requirements of tobacco tissue culture, *Physiol. Plant.*, 18, 100, 1965.
12. **Kasperbauer, M. J. and Reinert, R. A.,** Biological detection of phytochrome in callus tissue of *Nicotiana tabacum* L., *Nature (London)*, 211, 744, 1966.
13. **Kasperbauer, M. J. and Reinert, R. A.,** Photometrically assayable phytochrome in-vivo in callus tissue cultured from *Nicotiana tabacum, Physiol. Plant.*, 20, 977, 1967.
14. **Kasperbauer, M. J. and Collins, G. B.,** Reconstitution of diploids from leaf tissue of anther-derived haploids in tobacco, *Crop Sci.*, 12, 98, 1972.
15. **Kasperbauer, M. J. and Wilson, H. M.,** Haploid plant production and use, in *Nicotiana: Procedures for Experimental Use* (USDA Tech. Bull. 1586). Durbin, R. D., Ed., U.S. Department of Agriculture, Washington, DC, 1979, chap. 5.
16. **Quak, F.,** Meristem culture of virus-free plants, in *Plant Cell, Tissue and Organ Culture,* Reinert, J. and Bajaj, Y. P. S., Eds., Springer-Verlag, New York, 1977, 598.
17. **Clapham, D.,** Haploid *Hordeum* plants from anthers in vitro, *Z. Pflanzenzuecht.*, 69, 142, 1973.
18. **Wilson, H. M.,** Culture of whole barley spikes stimulates high frequencies of pollen calluses in individual anthers, *Plant Sci. Lett.*, 9, 233, 1977.

Chapter 5

HAPLOIDS: DERIVATION AND EVALUATION

M. J. Kasperbauer

TABLE OF CONTENTS

I. INTRODUCTION

When this author was in graduate school in the late 1950s, plant breeders could only dream that one day it would be possible to examine the genetic information in gametes, select the desired ones, and then combine the selected gametes to form a desired hybrid. It was apparent that such technology could be very beneficial for more rapid improvement of wind-pollinated, highly self-infertile species such as tall fescue (*Festuca arundinacea*). Considerable progress has been made since then. This chapter will be devoted to the development and application of concepts that have led to the production of haploid plants from immature gametes.

Theories discussed by Haberlandt[1] provided an early step in the evolution of plant biotechnology. He reasoned that each somatic cell in a plant contains the same genetic code and that somatic cells should be capable of developing into whole plants, if given the appropriate surroundings. However, technology was not sufficiently advanced to test the theories during his scientific career. Later, pioneering studies of plant tissue and cell culture by Philip White, R. J. Gautheret, F. C. Steward, Folke Skoog, and others provided a practical beginning to the *in vitro* culture of plant tissue and the regeneration of plants from the cultured tissue.

A report by Guha and Maheshwari,[2] in 1964, demonstrated *in vitro* production of embryos from anthers of *Datura innoxia*. A few years later, the same investigators[3,4] reported further success in development of embryoids from pollen grains of *Datura*. Within the next several years, a number of scientists, most notably Nitsch and colleagues,[5-8] were successful in culturing haploid tobacco (*Nicotiana tabacum*) plants from immature anthers. A paper by Nitsch and Nitsch,[7] in 1969, was followed by an "explosion of interest" in this approach and in numerous attempts to culture androgenetic haploid plants from a wide range of species. It is of significance that the most success was obtained with tobacco and a few other members of the *Solanaceae* family, which includes *Datura*.

During the mid 1960s, this author began collaborative research with Reinert using cultured tobacco tissue to learn that phytochrome was present in dividing and recently divided cultured cells, and that phytochrome played a role in uptake and metabolism by such tissue.[9-11] The finding that red light (acting through the phytochrome system within the cultured tissue) favored certain metabolic responses in cultured tissue provided an important basis for selection of light and temperature conditions that were later used in development of our anther culture and plant regeneration procedures.[12-14] As success with tobacco anther culture evolved,[12-16] Buckner persuaded them to join with him in a collaborative approach in his tall fescue breeding program. Our goal was to accomplish with tall fescue what was already being done with tobacco. That is, to culture haploid plants from immature anthers, to evaluate the haploid plants under controlled environments and field stress conditions, and to produce doubled haploids from the haploid plants that performed best in the evaluations.[17] Because tall fescue is wind pollinated and highly self-infertile, there is a high probability that each gamete (and gamete-derived haploid plant) may be genetically different from each other gamete (or haploid plant derived from another gamete). The remainder of this chapter is devoted to the development of theory and procedures that led to successful production and evaluation of cytologically verified haploid tall fescue plants from immature anthers.

II. DERIVATION OF HAPLOIDS

The objective was to cause the gametic genetic code to be expressed in the form of functional haploid tall fescue plants. After cloning such haploid plants via tillers or via tissue culture and plant regeneration (see Chapter 4 of this book), one should be able to evaluate them under a wide range of environmental conditions to identify the best genotypes. Chromosome numbers should then be doubled in the selected haploids to produce fertile pure lines with the characteristics that were selected during evaluation of the haploid plants. Derivation of tall fescue haploid plants can be approached by several different procedures.

A. Twin Seedlings

Haploid plants can occur as natural accidents. The occurrence of haploids via twin seedlings varies in frequency among genotypes, even within the same species. In a conventional forage grass breeding program, such haploid plants are usually lost because they are smaller than normal seed-borne plants and they are not fertile. However, they could be useful in study of chromosome pairing.[18]

B. Maternal Haploids

Maternal haploids have been derived in several crop plants. For example, haploid embryos develop on barley (*Hordeum vulgare*) plants after pollination by *H. bulbosum*.[19-22] After fertilization, the *H. bulbosum* chromosomes are eliminated during subsequent mitotic divisions, and the developing embryo is a haploid *H. vulgare*. Another example involves tobacco (*Nicotiana tabacum*) plants. The *N. tabacum* flowers are emasculated before pollen shed, and the stigmas are dusted with pollen from *N. africans* plants.[23] As with barley, the genetic contribution of the pollen parent is eliminated and gynogenetic haploid embryos develop.[23] Also, culture of some haploid plants from unfertilized ova should be possible, as is the culture of haploid plants from immature anthers.

Early experimental evidence suggests that gynogenetic haploids may be more vigorous than androgenetic haploids. This may be an important factor for future use of haploids in plant improvement. Certainly, comparisons of the two kinds of haploids should be made.

Because of the unique nature of the tobacco plant as a model system for comparison of the various approaches, it is possible to obtain anther-derived haploids, maternal haploids, and inbred seed from the same plant. To do this, (1) some floral buds can be excised to obtain immature anthers for culture, (2) some immature ova can be cultured, (3) anthers can be removed (before pollen shed) from other floral buds which are subsequently pollinated with pollen from *N. africans*, and (4) some of the flowers can be self-pollinated to produce inbred seed. Thus, the large numbers of microspores per anther, ova per flower, and flowers per plant should allow an investigator to obtain both androgenetic and gynogenetic haploids from the same tobacco plant. After doubling chromosome numbers in such haploids, the doubled haploids could be self-pollinated and progeny of the two kinds of doubled haploids could be compared with the inbred progeny of the same original source plant. Thus, tobacco can provide a valuable model system in which to compare the vigor and potential usefulness of anther-derived vs. maternal haploid plants of the same genotype. Currin and colleagues at Clemson University have initiated such studies.[24] These will undoubtedly provide an extremely useful model system for other crop plants, including tall fescue.

C. Anther-Derived Haploids

The pioneering research of Guha and Maheshwari[2-4] and that of Nitsch and Nitsch[7] led to considerable discussion of the potential value of anther-derived haploids in plant breeding.[15,25-27] High enthusiasm among plant breeders, geneticists, and administrators followed the early success with tobacco. However, culture of haploids from other species, especially the grasses, was much more difficult. For example, excised anthers had to be stimulated to first produce callus and then to regenerate plants from the callus in a two-step process with some cereal grasses including rice (*Oryza sativa*),[27] wheat (*Triticum aestivum*),[28] and barley.[29] Although some plants were regenerated from such callus, the total number was extremely low and frequently included albinos and various ploidy levels.[28] Using a modified approach, Wilson[30] obtained callus from barley spikes that were cultured in liquid medium which contained 10% sucrose. He regenerated about 40 plants from the callus, but only 2 of them were green and they soon died. Other investigators also found that callus which formed on high-sucrose media regenerated a higher percentage of albinos and other abnormal plants relative to callus formed on media with lower levels of sucrose.[31]

While early attempts to culture haploid grass plants from excised anthers were not very successful, some researchers were able to initiate and maintain callus cultures from somatic tissue of various members of the grass family.[32-27] Eventually, numerous green plants were regenerated from somatic tissue-derived callus cultures of forage grasses[34,35,37] (see also Chapter 4 of this book). Our attempts to obtain tall fescue anther-derived haploid plants via excised anthers and via anthers attached to a segment of panicle (nurse tissue) are discussed below.

1. Rationale and General Approach

Earlier experience with tobacco showed that it is possible to culture green plants from immature gametes in anthers, and that such plants have the same chromosome numbers as the gametes from which they originated.[5-8,14] The haploids have been useful for evaluation of gene expression in growing plants. Doubling the chromosome numbers of such haploids results in fertile, pure lines with the genetic characteristics that were selected in the original haploid plants.[13,15] These same capabilities with a highly heterozygous species such as tall fescue could be even more useful in accelerating development of superior breeding lines.

We used field-grown plants of a widely used cultivar, ''Kentucky 31'', as a source of tall fescue anthers. Field-grown source plants were preferred because they had demonstrated survival under field conditions. There were, however, some disadvantages to using field-grown plants as a source of anthers. Field-grown plants often contained an endophytic fungus (*Acremonium coenophialum*), and decontamination of the tissue was difficult. Tall fescue has the added disadvantage (relative to annuals such as tobacco, wheat, and barley) of being very seasonal. That is, this perennial species must go through a period of cold vernalization followed by long days to bring it into flower. A combination of the importance of tall fescue as a forage and turf grass, and the difficulties associated with its use in cell and tissue culture approaches, have resulted in a highly challenging area of research.

The tall fescue plants were selected (generally from vigorous field-grown plants in springtime) when the panicles showed about 2 to 4 cm above the flag leaf.[38] The plants were cut at the soil surface and immediately placed in bottles of tap water for transport to the laboratory. Clear plastic bags were then placed over the fescue panicles to avoid drying of the young tissue during the preculture period. The protected materials were placed in darkness at 5°C for various durations before the anthers were cultured. Beneficial effects of a ''cold shock'' treatment were discussed by Sunderland and Roberts.[39] In addition to possible benefits from the cold shock treatment, some flexibility in duration of storage time at 5°C allows collection of a large number of plants from field plots at the appropriate stage of development, with some flexibility to initiate the cultures. This flexibility allows more efficient use of laboratory facilities and personnel. We used cold storage durations ranging from 0 to more than 30 d before processing the materials and starting the cultures.

The first step in processing the material was to stage representative anthers and relate this to general appearance of the florets from which they came.[38] In this way, we were able to visually examine the panicles and to remove and discard spikelets with a high probability of having anthers with microspores that were either earlier or later than the uninucleate stage. The remaining portions of the panicles (with only the selected florets attached) were then surface decontaminated prior to culture of anthers. The panicles, with remaining spikelets, were immersed for 5 min in a 0.3% sodium hypochlorite solution to remove surface contamination. All remaining preculture steps were done with sterile instruments in a laminar-flow transfer hood. The panicles were then rinsed in three changes of sterile water to remove the sodium hypochlorite. No attempt was made to remove any contaminant that may have been present within the plant tissue because of the high probability of damaging the dividing plant cells. Both excised anther and anther-panicle cultures followed the same procedures from plant collection through surface decontamination of the selected portions of the panicles.

2. *Excised Anthers*

Anthers were excised from tall fescue panicles that were pretreated in various ways and they were cultured on various media that were suggested by previous investigators.[28,29,31,37] The basic medium was a slight revision of that used in the mid 1960s for tobacco.[10] It was modified from Linsmaier and Skoog[40] and from Murashige and Skoog.[41] The basic medium contained the inorganic salts and organic substances outlined in Chapter 4 of this book. Variables in culture media included 2 to 12% sucrose, 0.to 20 mg of 2,4-D per liter, 0 to 2 mg kinetin per liter, 0.0 to 2.5 g active charcoal per liter, and different levels of casein hydrolysate and inositol. Attempts were made to obtain haploid plants or callus from excised anthers of tall fescue plants of several cultivars and ecotypes from two different growing seasons.

a. *Observations*

No haploid plants or calli were obtained from any of the thousands of excised anthers in the initial studies with tall fescue. Even though all attempts with excised anthers failed, some of the same medium combinations and plant ecotypes were successful in development of callus from soft, young somatic tissue that was excised from the lower ends of peduncles and internodes. The results obtained with the young somatic tissue explants proved that the culture medium was capable of initiating and supporting callus growth from tall fescue. Nevertheless, fully expanded leaf and stem tissue from the same plants failed to form callus or to regenerate plants.

b. *Interpretation*

Several important concepts evolved from the combination of experiments with excised anthers, young somatic tissue, and fully expanded somatic tissue. It was apparent that the young stem tissue which contained many dividing and recently divided cells could continue to undergo cell division to form callus and regenerate plants,[37] whereas fully expanded somatic tissue would not form callus or plantlets when using our media and culture protocols.[37] It was also evident that the young somatic tissue which contained dividing or recently divided cells continued to undergo divisions to initiate callus and regenerate plantlets, while no callus initiation or plant regeneration occurred from the excised anthers, which also contained dividing cells at the time of excision. There were several possible interpretations for these apparently conflicting responses. Perhaps the young stem explants contained some growth factor, or factors, that was needed for continued cell division to occur on the media under the culture conditions used. If true, perhaps the factor was missing, or present in inadequate amounts, in the excised anthers (this might be visualized as a need for a "critical mass" of tissue).

3. *Anther-Panicle Nurse Culture*

a. *Theory*

The relative ease of initiating callus formation and plant regeneration from young somatic tissue explants, and the complete failure with excised anthers from the same plant and on the same medium formulations, led to the hypothesis that suitable nurse tissue might be needed to provide a stimulus to the microspores in the cultured anthers. Several different approaches to the use of nurse tissue and/or the addition of plant extracts to the culture medium were considered. The desired goal was to cause the microspores within the anthers to continue divisions and to then develop into haploid plants. To accomplish this, the nurse tissue and/or culture medium should contribute a stimulus to the dividing microspores, but the nurse tissue should not form any callus or regenerate any plants that would appear to be similar to those originating from the microspores within the anthers.

Our previous work with tall fescue and its hybrids (Chapter 4 of this book) provided the needed starting point for the approach with nurse tissue. The key point in the earlier study was that young somatic tissue which contained dividing or recently divided cells would continue to undergo cell division to form callus and regenerate plants on the media used, whereas fully expanded leaf and stem explants from the very same plants remained green in culture, but failed to develop any callus or regenerate any plants. This led to the hypothesis that fully expanded panicle tissue might be successful as nurse tissue if it remained attached to the appropriate stage of anthers. Such a combination would allow the dividing microspores in the anthers to remain attached to their natural nutrient source while also being affected by the components of the culture medium. Also, if the panicles were carefully selected and trimmed so that all somatic tissue capable of cell division (on the selected culture medium) was removed, there should be no callus formation or plant regeneration from somatic tissue, while callus and/or plantlets would develop from the immature gametes.

b. *Approach*

Vigorously growing plants were preselected for desirable agronomic characteristics, forage quality factors, and winterhardiness. Plant materials were collected from the field at about 0900 to 1000 h when the panicles began to show above the flag leaf. At that stage of panicle development, there was a wide range of anther stages on each panicle.

As discussed under Section II.C.1, above, the tall fescue stems with emerging panicles were covered with plastic bags to maintain humidity and to avoid drying during storage at 5°C prior to preparation for culture. This allowed us to collect many inflorescences from field plants over a relatively short time period when plants were at the appropriate stage of growth and to then store them for up to several weeks prior to processing. We used cold storage preculture periods ranging from 0 to more than 30 d in the various experiments.

Special care was taken to avoid drying of the anthers during processing and placement on culture medium. This was accomplished by keeping some sterile water over the anthers in the dissecting dish while they (or small panicle segments) were handled prior to placement on the appropriate culture media. After a few preliminary experiments (discussed under Section II.C.3.c, below), preparation of panicles for the anther-panicle culture approach involved (1) identification and removal of florets with anthers having a high probability of being younger than the uninucleate stage (based on microscopic examination of similar material), and (2) removal of all panicle tissue from which the young florets were already removed. The desired goal was to remove all somatic tissue capable of callus initiation and/or plantlet regeneration under the culture conditions. If totally succesful, the somatic tissue would serve as nurse tissue and only the immature gametes would undergo cell divisions and develop into plants.

Nutrient medium composition is very important and should be carefully planned to give desired results. For example, the exogenous hormone balance should be such that it favors cell divisions followed by haploid plantlet regeneration. Use of a multiple-step approach seems logical. Our experience with various tissue culture approaches has shown that preliminary experiments are often useful to determine genotype responsiveness to the level of auxin in the medium. While high amounts of 2,4-D will frequently cause initiation of more callus, lower amounts result in a much lower frequency of abnormal callus and a lower frequency of abnormal plants among those regenerated from the callus tissue.

Tall fescue appears to be autonomous for cytokinin. That is, the balance between callus initiation and plant regeneration can be manipulated by merely adjusting the auxin level in the medium.[37] This is an important difference from tobacco in which the balance between callus growth and plant regeneration is manipulated by altering the exogenous auxin/cytokinin ratio. Tobacco callus can be caused to regenerate plants by decreasing the auxin/cytokinin ratio. This can be accomplished by either decreasing the auxin and leaving cytokinin constant or by increasing the cytokinin and leaving auxin constant.[16] With tall fescue and some other plants that are autonomous for cytokinin, a concentration of about 2 mg of 2,4-D per liter of culture medium is usually adequate to initiate callus formation from young somatic tissue.[37] Decreasing the concentration of 2,4-D to about 0.25 mg per liter, or even less, then has the effect of decreasing the auxin/cytokinin ratio, which results in a decrease in new callus formation and an increase in plant regeneration from the callus.

Sugar is another medium component that can influence plant regeneration from callus.[29,30] With tall fescue somatic tissue, we have found that sucrose concentrations of 10 to 12% result in regeneration of more plants than are obtained with 2 to 3% concentrations. However, the greater number of regenerated plants associated with the higher sucrose concentrations in the medium also includes higher percentages of abnormal plants such as albinos.

To be successful with the tall fescue anther-panicle approach requires (1) very careful removal of all undesired ("young") somatic tissue, (2) care to avoid drying the fragile young anthers, and (3) use of suitable media with carefully controlled manipulation of the auxin level to first initiate embyrogenic callus from the microspores and to then regenerate haploid plantlets. The success of some of these manipulations can be influenced by factors as subtle as medium preparation and relatively minor temperature fluctuations during culture. Our earlier studies of phytochrome involvement in uptake and metabolism in tobacco callus[10] suggest that the light conditions of the culture room might be expected to influence responses to hormone level in the medium.

As those familiar with tissue culture and plant regeneration realize, there is frequently a fine line between success and failure. Therefore, one should first do some preliminary experiments to become familiar with the materials at hand and then take all precautions to select or develop protocols with the highest probability of success.

c. Results and Discussion

Our experiments with anther-panicle culture evolved through several stages. The preliminary attempts involved merely surface decontamination of the panicles, cutting them into 2- to 3-cm segments and placing them in petri dishes on the culture medium previously used successfully for "young" stem tissue,[37] and unsuccessfully for excised anthers of tall fescue.[38] No segments of the panicles were discarded prior to culture; therefore, some segments contained "young" somatic tissue, others only "aged" tissue, and still others were a mixture. The first visually detectable changes in the cultured panicle segments occurred in about 10 d. Unfortunately, it soon became apparent that callus formation was originating from young somatic tissue in some of the panicle pieces. On the other hand, these preliminary experiments offered encouragement because some of the segments appeared to initiate callus only from anthers and not from the attached somatic tissue. Other segments (from the older portions of panicles) did not produce any callus from either anthers or panicle tissue. The preliminary experiment was terminated when it became apparent that callus was forming on some of the "young" panicle tissue.

The interpretation of the preliminary anther-panicle experiment was that:

(1) the culture medium was suitable to initiate callus, (2) young somatic tissue (that capable of callus formation) should be removed before culture, and (3) a two-stage procedure should be used to culture the appropriate stage of anthers while attached to nurse tissue that was not capable of callus formation. The goal was to initiate embryogenic callus from only the microspore cells, and then transfer to fresh medium with a lower 2,4-D concentration to regenerate plantlets from the newly initiated callus.

The first attempt with anther-panicle culture following the interpretation outlined above resulted in successful derivation of haploid plants. Field-grown "Kentucky 31" tall fescue plants were collected when the panicles showed about 3 cm above the flag leaf. After transfer to the laboratory, the shoots were covered with plastic bags and held in darkness at 5°C for more than 3 weeks, while other materials from the same collection were used in various preliminary experiments. Based on those preliminary experiments, all spikelets with anthers earlier or later than the uninucleate stage were discarded (stage of anthers was estimated by visual appearance based on cytological examinations of similar materials). We also discarded the panicle segments from which young anthers were already removed. The panicles, with remainig spikelets, were then surface decontaminated for 5 min in 0.3% sodium hypochlorite. The decontaminated panicles were then rinsed in three changes of sterile water to remove the decontamination solution. Each panicle was then placed in a separate petri dish containing some sterile water to keep the anthers from drying during the preculture preparation process. The panicles were then cut into 2- to 3-cm segments and placed in petri dishes on appropriate culture media for the first stage. We used petri dishes because they were convenient to handle and to stack during culture. This was especially convenient if some of the panicle segments had to be "rescued" because others in the same dish contained an endophytic contaminant that became apparent during culture. The culture room was kept at 22 ± 1°C in darkness, except for observation periods that varied from several times per week to several times per day. It should be noted that such a fequency of exposure to low-intensity light could influence involvement of the phytochrome system, without introducing photodegradation of media components as could occur with prolonged exposure to intense light.

The initial culture medium was modified from the one used earlier for tobacco,[10] but with the auxin content used for tall fescue callus initiation as shown in Chapter 4 of this book. Details are reported here for convenience. Each liter of medium contained *inorganic salts*: 1.65 g NH_4NO_3, 1.90 g KNO_3, 0.44 g $CaCl_2 \cdot 2\,H_2O$, 0.37 g $MgSO_4 \cdot 7\,H_2O$, 0.17 g KH_2PO_4, 15.9 mg $Fe_2(SO_4)_3$, 32.0 mg EDTA, 6.2 mg H_3BO_3, 22.3 mg $MnSO_4 \cdot H_2O$, 8.6 mg $ZnSO_4 \cdot 7\,H_2O$, 0.83 mg KI, 0.25 mg $Na_2MoO_4 \cdot 2\,H_2O$, 0.025 mg $CoCl_2 \cdot 6\,H_2O$, and 0.025 mg $CuSO_4 \cdot 5\,H_2O$); *organic substances:* 100 mg *myo*-inositol, 0.1 mg thiamine·HCl, 0.5 mg nicotinic acid, 0.5 mg pyridoxine·HCl, 2.0 mg glycine, 20 g sucrose, and 6 g agar; 2 mg 2,4-dichlorophenoxyacetic acid

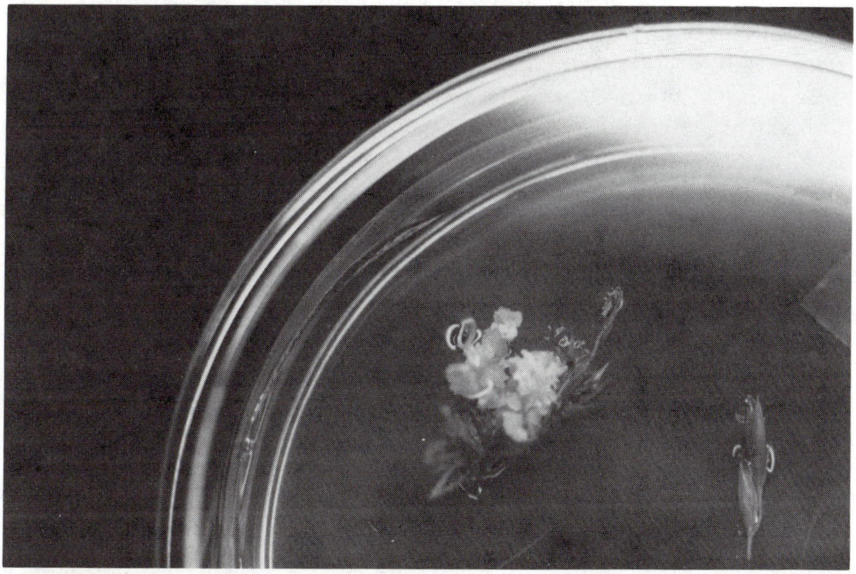

FIGURE 1. Plantlets and loose proliferation of cells from second stage of anther-panicle cultures of tall fescue. (From Kasperbauer, M. J., Buckner, R. C., and Springer, W. D., *Crop Sci.*, 20, 103, 1980.)

(2,4-D) was used for the first stage of the anther-panicle cultures. All components except sucrose, agar, and inositol were prepared as stock solutions, as described in Chapter 4 of this book. The pH was adjusted to 5.7 with 2 *N* NaOH after all components were added and the agar was melted. All media were autoclaved at 121°C for 15 min. About 50 ml of medium was added to each 100 × 20-mm sterile plastic petri dish and allowed to cool and stand for at least 1 d before use.

The second-stage medium was the same as the first-stage except that it had a lower 2,4-D concentration (i.e., 0.25 mg of 2,4-D per liter of medium) in order to cause the newly initiated callus to regenerate plants. The objectives were to stimulate embryogenic callus initiation from the uninucleate stage cells in the immature anthers during the first stage, and to then cause regeneration of haploid plants during the second stage. In this experiment, the materials were transferred from the first-stage medium (callus initiation) to the second-stage medium (plant regeneration) soon after the loose proliferation of cells became visually detectable.

A loose proliferation of cells with some plantlets emerging is shown in Figure 1. The first evidence of plantlet formation was a large number of tiny green flecks scattered through the proliferation of cells (because of the limited amount of material, none of it was sacrificed for microscopic study of embryoid formation). As plantlets began to develop, they were transferred to a

FIGURE 2. Haploid tall fescue plants that were derived via anther-panicle culture. (From Kasperbauer, M. J., Buckner, R. C., and Springer, W. D., *Crop Sci.*, 20, 103, 1980.)

rooting medium (half strength of the same basic formulation, but with no 2,4-D, and with sucrose and agar concentrations of 10 and 5 g/l, respectively). There were no albinos among the regenerated plantlets. More than 30 green plantlets that originated from different areas of the proliferation of cells were transferred to the rooting medium and placed in a growth chamber at 22 ± 1°C with 14-h daily light periods from cool white fluorescent lamps at about 80 μmol m^{-2} s^{-1}. After the plantlets developed roots, they were transferred to potting soil in a 20°C growth chamber and covered with tranparent plastic to maintain humidity during plant establishment. After several days, the plastic canopy was removed for about an hour twice per day to condition the plants. After the conditioning period, the plants received 14-h light periods from cool white fluorescent lamps at fluence rates that were increased over several days to about 550 μmols m^{-2} s^{-1}. Twenty-three of the plants survived the transfer to the large growth chamber in which they grew rapidly. Two representative plants are shown in Figure 2.

The chromosome number of each of the plants was determined by examination of root and leaf meristem metaphases by Springer. Details are discussed in Chapter 7 of this book. Chromosome numbers in somatic metaphase cells showed that all but one of the plantlets had the haploid number of chromosomes, n = 21 (Table 1, Figure 3). Some pronounced phenotypic differences existed among the various haploid lines, which supported the

TABLE 1
Chromosome Numbers in Somatic Metaphase Cells from Shoot Tips of Tall Fescue Plantlets Derived by Anther-Panicle Culture

Plant	Cells counted	Chromosome no.
1	27	21
2	50	39
3	34	21
4	25	21
5	42	21
6	30	21
7	25	21
8	35	21
9	25	21
10	6	21
11	27	21
12	25	21
13	25	21
14	25	21
15	25	21
16	31	21
17	25	21
18	29	21
19	25	21
20	25	21
21	15	21
22	25	21
23	50	21

From Kasperbauer, M. J., Buckner, R. C., and Springer, W. D., *Crop Sci.*, 20, 103, 1980.

concept that the different haploid plants originated from different microspores from a heterozygous ''parent'' plant. After chromosome counts were completed, the plants were grown under 12-h days to encourage tillering (as discussed in Chapter 4 of this book). After cloning the various haploid lines, ramets were transplanted to replicated field plots to evaluate them for winterhardiness, growth, development, and selected forage characteristics. Phenotypically, the 22 haploid lines ranged from early flowering with little leaf growth to sparse flowering with abundant leaf growth.[38] Some comparisons are shown in Table 2. Sugar, total nitrogen, moisture, dry matter disappearance, neutral detergent fiber (NDF), and phosphorous levels were used to estimate quality in summer and fall forage. Significant differences among lines occurred for all analyses. For example, NDF and sugar levels ranged from 48 to 59%, and 5.0 to 7.8%, respectively, for the summer harvest.[42] After field evaluation of the haploid lines, some of them were selected for development of doubled haploid lines, as discussed in Chapter 6 of this book.

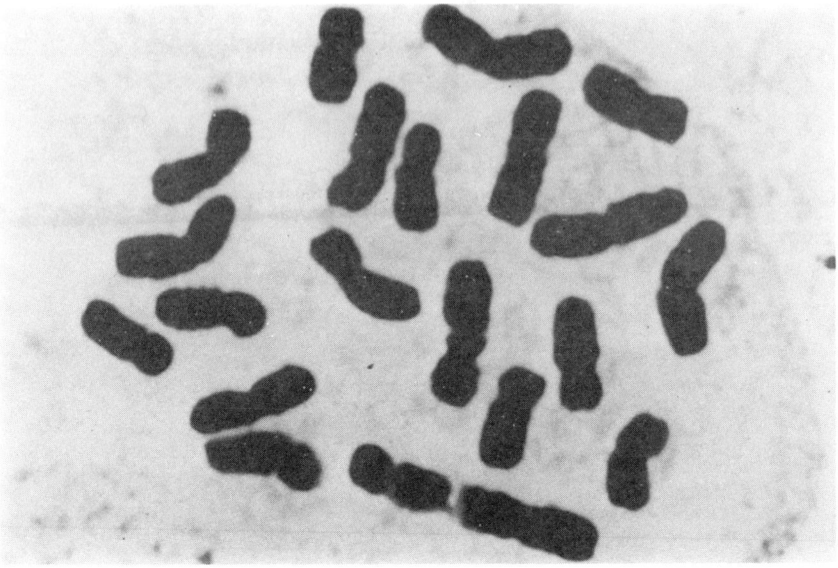

FIGURE 3. Chromosomes in shoot tip of a haploid tall fescue plant, n = 21. (From Kasperbauer, M. J., Buckner, R. C., and Springer, W. D., *Crop Sci.*, 20, 103, 1980.)

Subsequent experiments involving anther-panicle culture involved such variables as duration of preculture cold treatment, sucrose concentration of the medium, and 2,4-D levels during the first stage of anther-panicle culture. Plantlet regeneration was successful from a number of preculture cold treatments and culture medium combinations used in the first stage of culture. As suggested above, fewer abnormal plants were regenerated from anther-panicle segments that were cultured on the lower levels of 2,4-D and sucrose during the first stage. There were phenotypic variations among the normal-appearing regenerated plants, as would be expected among haploid plants that originated from different immature gametes of a wind-pollinated, self-infertile plant such as tall fescue. Because of the large number of regenerated plants and scheduling difficulties, cytological evaluations were limited to confirmation that randomly selected plants were haploids.

III. EVALUATION

A. Visual

Visual evaluation is suitable to identify and eliminate obviously abnormal plants. This can begin soon after plant regeneration by eliminating albinos and those that develop obviously abnormal roots and shoots. After establishment, additional visually detectable abnormalities may appear during greenhouse and field evaluations. Also, there may be visually detectable superior characteristics such as leafiness, color, and plant form.

TABLE 2
Relative Amounts of Vegetative and Reproductive Growth of the Different Androgenetic Haploid Lines Under Field Conditions near Lexington, KY, in 1980—1981

Haploid line[a]	Field growth			
	Veget.	Reprod.[b]	Dry matter (g dry wt/5 plants)	
	22 July		25 Aug.	28 Oct.
1	8	6	54	100
3	2	10	15	27
4	6	6	41	55
5	10	3	55	91
6	8	4	59	77
7	8	1	51	75
8	8	4	52	84
9	9	4	52	92
10	9	4	59	101
11	10	4	76	104
12	6	6	26	44
13	2	4	7	7
14	9	3	58	134
15	10	3	60	111
16	10	4	58	132
17	4	6	18	31
18	10	5	69	126
19	8	6	66	111
20	10	4	59	111
21	1	10	—	—
22	9	6	71	103
23	8	5	52	92
LSD 0.05			16	32

[a] Haploid line numbers are those used in the paper by Kasperbauer et al.[38] Plant no. 2 was not a haploid; therefore, it is deleted from this table.
[b] Rating scale is 1 to 10. 1 = least, 10 = most.

From Kasperbauer, M. J. and Eizenga, G. C., *Crop Sci.*, 25, 1091, 1985.

B. Cytological

Cytological evaluation of putative haploids is an important part of the evaluation process. Confirmation of haploidy is needed before one attempts to develop doubled haploids. Details of the importance of and procedures for cytological evaluation of regenerated tall fescue plantlets are outlined by Springer in Chapter 7 of this book.

C. Field

Regenerated plants that advanced through preliminary visual and cytological evaluations can be increased through tillers for further detailed evaluation. After cloning the material, ramets can be used in replicated field (or controlled environment) studies to determine winterhardiness, drought tolerance, forage quantity, forage composition, disease resistance, etc. After field evaluation, selected haploid plants might be used in the development of fertile doubled haploid, pure lines with the same genetic code that was present in the "parent" haploid plant.[42]

IV. POTENTIAL SIGNIFICANCE

The experience with anther-panicle cultures clearly showed that it is possible to culture haploid plants from immature gametes of tall fescue. The haploid plants can easily be cloned via tillers or through cell culture and plant regeneration. Theoretically, use of the cloned haploid plants allows evaluation of the gametic genetic code under a wide range of field and other stress environments to identify those genetic combinations (haploid plants) that have the most desirable characteristics and to eliminate the undesirable ones. Doubling the chromosome numbers in the selected haploids should result in fertile pure lines that can then be used in hybrids or in synthetics. The doubled haploids could be crossed with other doubled haploids or they could be crossed with lines derived through conventional tall fescue breeding approaches. Another potential use of cytologically verified tall fescue haploids is as the female parent in the development of monosomic lines for basic genetic studies.[43]

REFERENCES

1. **Krikorian, A. D. and Berquam, D. L.,** Plant cell and tissue cultures: the role of Haberlandt, *Bot. Rev.*, 35, 59, 1969.
2. **Guha, S. and Maheshwari, S. C.,** In vitro production of embryos from anthers of *Datura, Nature, (London)*, 204, 497, 1964.
3. **Guha, S. and Maheshwari, S. C.,** Cell division and differentiation of embryos in pollen grains of *Datura* in vitro, *Nature (London)*, 212, 97, 1966.
4. **Guha, S. and Maheshwari, S. C.,** Development of embryoids from pollen grains of *Datura* in vitro, *Phytomorphology*, 17, 454, 1967.
5. **Bourgin, J. P. and Nitsch, J. P.,** Obtention de *Nicotiana* haploides a partir d'etamines cultivees in vitro, *Ann. Physiol. Veg.*, 9, 377, 1967.
6. **Nitsch, J. P.,** Experimental androgenesis in *Nicotiana, Phytomorphology*, 19, 389, 1969.
7. **Nitsch, J. P. and Nitsch, C.,** Haploid plants from pollen grains, *Science*, 163, 85, 1969.
8. **Nitsch, J. P.,** Haploid plants from pollen, *Z. Pflanzenzuecht.*, 67, 3, 1972.
9. **Kasperbauer, M. J. and Reinert, R. A.,** Biological detection of phytochrome in callus tissue of *Nicotiana tabacum* L., *Nature (London)*, 211, 744, 1966.
10. **Kasperbauer, M. J. and Reinert, R. A.,** Photometrically assayable phytochrome in-vivo in callus tissue cultured from *Nicotiana tabacum, Physiol. Plant.*, 20, 977, 1967.

11. **Reinert, R. A. and Kasperbauer, M. J.,** Influence of red and far-red light on virus content of callus tissues cultured from *Nicotiana tabacum, Phytopathology,* 56, 1108, 1966.
12. **Kasperbauer, M. J. and Collins, G. B.,** Anther-derived haploids and reconstituted diploids in tobacco: techniques and evaluation, *Agron. Abstr.,* 8, 1973.
13. **Kasperbauer, M. J. and Collins, G. B.,** Reconstitution of diploids from leaf tissue of anther-derived haploids in tobacco, *Crop Sci.,* 12, 98, 1972.
14. **Kasperbauer, M. J. and Collins, G. B.,** Anther-derived haploids in tobacco: evaluation of procedures, *Crop Sci.,* 14, 305, 1974.
15. **Collins, G. B., Legg, P. D., and Kasperbauer, M. J.,** Use of anther-derived haploids in *Nicotiana.* I. Isolation of breeding lines differing in total alkaloid content, *Crop Sci.,* 14, 77, 1974.
16. **Kasperbauer, M. J. and Wilson, H. M.,** Haploid plant production and use, in *Nicotiana: Procedures for Experimental Use* USDA Tech. Bull. 1586, Durbin, R. D., Ed., U.S. Department of Agriculture, Washington, DC, 1979, chap. 5.
17. **Kasperbauer, M. J.,** Fescue cell and tissue culture (the potential to become an important part of forage grass improvement programs), in *Proc. of the 25th Grass Breeders Work Planning Conf.,* Barker, R., Ed., Northern Great Plains Research Center, Mandan, ND, 1979, 6.
18. **Malik, C. P. and Tripathi, R. C.,** Mode of chromosome pairing in the polyhaploid tall fescue (*Festuca arundinacea* Shreb 2n = 42), *Z. Biol.,* 116, 332, 1970.
19. **Kasha, K. J. and Kao, K. N.,** High frequency haploid production in barley (*H. vulgare* L.), *Nature (London),* 225, 874, 1970.
20. **Lange, W.,** Crosses between *Hordeum vulgare* and *H. bulbosum.* I. Production, morphology and meiosis of hybrids, haploids and dihaploids, *Euphytica,* 20, 14, 1971.
21. **Lange, W.,** Crosses between *Hordeum vulgare* and *H. bulbosum.* II. Elimination of chromosomes in hybrid tissues, *Euphytica,* 20, 181, 1971.
22. **Kasha, K. J. and Sadasivaiah, R. S.,** Genome relationships between *Hordeum vulgare* and *H. bulbosum, Chromosoma,* 35, 264, 1971.
23. **Burk, L. G., Gerstel, D. V., and Wernsmann, E. A.,** Maternal haploids of *Nicotiana tabacum* L. from seed, *Science,* 206, 585, 1979.
24. **Currin, R. E.,** Personal communication, 1988.
25. **Melchers, G.,** Haploid higher plants for plant breeding, *Z. Pflanzenzuecht.,*67, 19, 1972.
26. **Nickell, L. G. and Heinz, D. J.,** Potential of cell and tissue culture techniques as aids in economic plant improvement, in *Genes, Enzymes, and Populations,* Srb, A. M., Ed., Plenum Press, New York, 1974, 109.
27. **Oono, K.,** Production of haploid plants of rice (*Oryza sativa*) by anther culture and their use for breeding, *Bull. Natl. Inst. Agric. Sci. Ser. D (Tokyo).* 26, 139, 1975.
28. **Ouyang, T., Hu, H., Chuang, C., and Tseng, C.,** Induction of pollen plants from anthers of *Triticum aestivum* L. cultured in vitro, *Sci. Sin.,* 16, 79, 1973.
29. **Clapham, D.,** Haploid *Hordeum* plants from anthers in vitro, *Z. Pflanzenzuecht.,* 69, 142, 1973.
30. **Wilson, H. M.,** Culture of whole barley spikes stimulates high frequencies of pollen calluses in individual anthers, *Plant Sci. Lett.,* 9, 233, 1977.
31. **Zenkteler, M. and Misiura, E.,** Induction of androgenic embryos from cultured anthers of *Hordeum, Secale,* and *Festuca, Biochem. Physiol. Pflanz.,* 165, 337, 1974.
32. **Carter, O., Yamada, Y., and Takahashi, E.,** Tissue culture of oats, *Nature (London),* 214, 1029, 1967.
33. **Cheng, T. Y. and Smith, H. H.,** Organogenesis from callus culture of *Hordeum vulgare, Planta,* 123, 307, 1975.
34. **Kasperbauer, M. J., Buckner, R. C., and Bush, L. P.,** Plantlet regeneration from stem-derived tissue cultures of *Lolium multiflorum* × *Festuca arundinacea* hybrids, *Agron. Abstr.,* 61, 1977.

35. **Conger, B. V. and Carabia, J. V.,** Callus induction and plantlet regeneration in orchard grass, *Crop Sci.,* 18, 157, 1978.
36. **Green, C. E., Phillips, R. L., and Kleese, R. A.,** Tissue cultures of maize (*Zea mays* L.): initiation, maintenance, and organic growth factors, *Crop Sci.,* 14, 54, 1974.
37. **Kasperbauer, M. J., Buckner, R. C., and Bush, L. P.,** Tissue culture of annual ryegrass × tall fescue hybrids: callus establishment and plant regeneration, *Crop Sci.,* 19, 457, 1979.
38. **Kasperbauer, M. J., Buckner, R. C., and Springer, W. D.,** Haploid plants by anther-panicle culture of tall fescue, *Crop Sci.,* 20, 103, 1980.
39. **Sunderland, N. and Roberts, M.,** New approach to pollen culture, *Nature (London),* 270, 236, 1977.
40. **Linsmaier, E. M. and Skoog, F.,** Organic growth factor requirements of tobacco tissue culture, *Physiol. Plant.,* 18, 100, 1965.
41. **Murashige, T. and Skoog, F.,** A revised medium for rapid growth and bioassays with tobacco tissue cultures, *Physiol. Plant.,* 15, 473, 1962.
42. **Kasperbauer, M. J. and Eizenga, G. C.,** Tall fescue doubled haploids via tissue culture and plant regeneration, *Crop Sci.,* 25, 1091, 1985.
43. **Eizenga, G. C. and Kasperbauer, M. J.,** Chromosome pairing in tall fescue haploids derived by anther-panicle culture, *J. Hered.,* 76, 99, 1985.

Chapter 6

DOUBLED HAPLOIDS: DERIVATION AND EVALUATION

M. J. Kasperbauer

TABLE OF CONTENTS

I. INTRODUCTION

The primary purpose for doubling the chromosome numbers in a haploid tall fescue (*Festuca arundinacea* Shreb.) plant is to obtain a pure, fertile line with the genetic characteristics of the "parent" haploid plant. Doubled haploid plants should be the genetic equivalents of highly inbred lines, if no mutations occur during any of the chromosome doubling and plant regeneration steps. Therefore, use of haploids and doubled haploids offers tremendous promise for more rapid genetic improvement of wind-pollinated, highly self-infertile grass plants such as tall fescue.

Genetic improvement of tall fescue by conventional procedures is extremely slow.[1] Because of the need for inbreds (or their equivalent) and the natural barriers to inbreeding in conventional grass breeding approaches, we initiated research toward derivation and use of tall fescue haploids and doubled haploids.[2] The procedures used to derive tall fescue haploids are discussed in Chapter 5 of this book. Some of the haploid lines that displayed visually detectable "markers" and/or specific desirable agronomic characteristics[3] were used in experiments designed to obtain fertile doubled haploid lines with the markers and/or agronomic characteristics that were identified in the "parent" haploid plants.[2]

There is a high probability that a large amount of genetic variability exists among gametes in a wind-pollinated, highly self-infertile species such as tall fescue. Therefore, it would be desirable to examine expression of the genetic code in the form of haploid plants under various environmental conditions to eliminate undesirable ones and to select the most promising ones. This procedure would have the effect of allowing identification of desirable genotypes under various field and controlled stress environments. However, haploid tall fescue plants are male sterile and express only a low percentage of female fertility.[4] The objective of this chapter is to discuss theory and approaches used to develop fertile doubled haploid plants from cytologically verified, gamete-derived haploid plants of tall fescue.

II. DERIVATION

After field testing a number of the tall fescue haploid lines (see Chapter 5 of this book), the next goal was to obtain fertile doubled haploid lines from some of the haploids. Two approaches were used. They were (1) conventional colchicine treatment, as described by Buckner et al.,[5] and (2) a new tissue culture procedure which involved induced endomitotic divisions followed by regeneration of doubled haploid plants from the cells with the doubled chromosome numbers.[2]

Some of the cytologically verified haploids that were cloned via tillers and evaluated under field conditions (see Chapter 5 of this book) were used in attempts to obtain doubled haploids.[2] Five clumps of each of six different haploid lines were transferred from the field nursery (near Lexington, KY) to a greenhouse in mid February, 1981, after natural vernalization had occurred.[2] Each clump was put in a separate clay pot. Two clumps of each inbred line were treated with colchicine on March 2. Two more clumps of each of the six haploid lines were treated with colchicine on March 10. The other clump of each of the six haploids served as an untreated control and as the source of somatic tissue for the tissue culture approach to doubling chromosome numbers. After colchicine treatment, all plants, including the untreated controls, were grown at 18 ± 2°C in a greenhouse. They received supplemental light from cool white fluorescent lamps at 75 μmol m^{-2} s^{-1} to provide 16-h daily photoperiods. Long photoperiods in the greenhouse after the natural vernalization (which occurred in the field before mid February) caused plants to develop inflorescences.

A. Colchicine

Plant growth, preconditioning, and colchicine treatments were very similar to procedures used by Buckner et al.[5] in the doubling of chromosome numbers in the 28-chromosome hybrid of annual ryegrass (*Lolium multiflorum* Lam.) × tall fescue. Although we followed the previously described protocols, and the process was repeated with a second set of haploid plants and a fresh source of colchicine, no doubled haploid tall fescue plants were obtained by this method.[2] Essentially the same procedure was repeated the following year and, again, no doubled haploids resulted.

The colchicine-treated plants were grown in the long-day greenhouse and observed for pollen-shedding anthers after the panicles emerged. After the panicles dried, they were harvested to determine whether any seed developed. No pollen shed was observed and no seed developed on any of the colchicine-treated plants. It should be noted, however, that greenhouse conditions might have affected pollen development and that failure to set seed could have been due in part to a limited amount of pollen available from other tall fescue plants in the greenhouse. Nevertheless, even if seed had developed on the colchicine-treated plants, there was a high probability that the seed would have been hybrid, because tall fescue is wind-pollinated and highly self-infertile. This would have been inconsistent with our goal to obtain doubled haploid plants that could be cytologically verified, vegetatively increased (cloned), and evaluated under field and other stress conditions before introducing them into a breeding program as pure lines for certain selected characteristics.

Although the objective was to compare doubled haploids derived by the colchicine procedure with those derived by alternate procedures, we abandoned the colchicine approach.[2] At that point, we decided to pursue a tissue culture approach patterned after a procedure that was successful for tobacco (*Nicotiana tabacum* L.).[6]

B. Tissue Culture

1. Development of Rationale

As in many tissue culture approaches, previous research with tobacco served as the model system.[6] Doubled haploids of tobacco can be regenerated quite easily from somatic tissue taken from fully expanded leaves. In the tobacco model system, midvein tissue is excised from a fully expanded leaf of a cytologically verified haploid plant, surface decontaminated, and placed on a solidified culture medium with a tenfold decrease in the auxin/cytokinin ratio (usually as a tenfold increase in cytokinin concentration).[6] Numerous shoots develop directly on the cultured tissue within 3 to 4 weeks. An important point here is the observation that nearly all of the plants regenerated from midveins of expanding (''young'') leaves (from a cytologically verified haploid plant) are haploid, whereas many doubled haploid plants are regenerated from midveins of fully expanded (''aged'') leaves of the same haploid ''parent'' plant.[6]

The developmental stage (''young'' vs. ''aged'') of the somatic tissue explant appears to influence the frequency of endomitotic divisions prior to regeneration of plants. This concept was later discussed at the 1974 International Symposium on Haploids in Higher Plants.[7] There appears to be relationship between regeneration of doubled haploid plants and the fact that nearly all cell divisions had stopped in the fully expanded (''aged'') leaf tissue.

Although this procedure is very successful for tobacco, considerable modification was needed for tall fescue. Earlier experiments with tall fescue and its hybrid had already revealed that somatic tissue from fully expanded leaves or stems would not form callus, nor would the tissue regenerate plants.[8] In fact, only ''young'' somatic tissue with numerous dividing or recently divided cells was capable of forming callus on the nutrient media and with protocols used.[8] In that previous research, abundant callus could be initiated from ''young'' somatic tissue that was excised from the base of rapidly elongating peduncles and internodes, and the callus could be maintained through repeated subcultures on nutrient medium containing 2 mg of 2,4-D per liter.[8] Numerous shoots could be regenerated by placing the callus on the same basic medium formulation with the 2,4-D concentration decreased to 0.25 mg, or less, per liter. If the shoots were regenerated after one or two subcultures of 3 or 4 weeks of duration, nearly all of the regenerated plants had the same chromosome number as the ''parent'' plant. However, if there were many subcultures on maintenance medium before regenerating shoots, there were often many somaclonal variants among the regenerated plants. When shoots were regenerated after callus establishment and one prolonged subculture on maintenance medium, some of the regenerated plants had increased chromosome numbers relative to the ''parent'' plant.[8] These combined results from the earlier tobacco and tall fescue experiments provided the basis for successful derivation of tall fescue doubled haploids via tissue culture.[2]

TALL FESCUE

↓

Cytologically-verified haploid

↓

Excise ('young') stem segment

↓

Initiate callus

↓

Excise callus from primary plant explant, after 4 weeks

↓

Sub-culture once and 'age' callus until it begins to senesce

↓

Regenerate shoots

↓

Excise shoots

↓

Develop roots

↓

Cytologically verify ploidy in root tips

↓

Grow and evaluate doubled haploids

TOBACCO

↓

Cytologically-verified haploid

↓

Excise midvein from fully-expanded ('aged') leaf

↓

Regenerate shoots

↓

Excise shoots

↓

Develop roots

↓

Cytologically verify ploidy in root tips

↓

Grow and evaluate doubled haploid

FIGURE 1. Flow chart comparing the regeneration of doubled haploid shoots from somatic tissue excised from cytologically verified tall fescue haploids with procedures used for tobacco. (From Kasperbauer, M. J. and Eizenga, G. C., *Crop Sci.*, 25, 1091, 1985.)

2. Procedure

A tentative flow chart (Figure 1) was developed for regenerating doubled haploid tall fescue shoots after considering both the tobacco model system and the previous experiments with tall fescue and its hybrid. The goal was to induce endomitotic divisions in "aged" somatic tissue and then regenerate doubled haploid plants from the cells with the doubled chromosome number. After considering the earlier failure to initiate callus or regenerate plants from fully expanded ("aged") leaves and stems of tall fescue and the successful regeneration of tobacco doubled haploid shoots from "aged" leaf tissue, we proposed induction of haploid callus from "young" tall fescue stem tissue followed by "aging" the callus before regenerating plants. Further, we were influenced by the previous observation that an increased number of subcultures (prior to regeneration of shoots) would increase the frequency of aneuploids and other somaclonal variants.[8] Considering all of these factors, we hypothesized that the starting point for tall fescue should be establishment of callus from "young" somatic tissue taken from the lower end of rapidly elongating peduncles (or internodes) of cytologically verified haploid plants.[2] We used relatively low levels of sucrose and 2,4-D (20 g and 2 mg/l, respectively) in the callus establishment and maintenance medium because earlier experiments with other grass species produced more albinos and other abnormal plants when higher sucrose concentrations were used.[9,10] Our own experience (Chapter 5 of this book) was consistent with those results.

The soft "young" tissue was excised from the base of rapidly elongating peduncles and internodes of cytologically verified haploid plants that were sampled just as the panicles were beginning to emerge above the flag leaf. The explants were surface decontaminated, cut into 2- to 3-mm segments, and placed on basic callus establishment and maintenance medium (see Table 4 in Chapter 4 of this book) in 20 × 100-mm sterile petri dishes. The cultures were kept in continuous light (15 μmol m^{-2} s^{-1}) from cool white fluorescent lamps in a windowless laboratory that was kept at 23 ± 2°C. After about 4 weeks, the callus was excised from the individual explants and transferred to fresh medium of the same formulation (with 2 mg 2,4-D per liter). The relatively small (only a few millimeters in size) pieces of callus were left on the fresh medium for only one extended subculture before transfer to the shoot regeneration (low 2,4-D) medium (see Table 4 in Chapter 4 of this book). The rationale was that rapid callus growth should occur during the first 3 or 4 weeks followed by "aging" of the callus. As predicted, the calli grew rapidly during the first several weeks, then the growth rate slowed and many of the calli had begun to senesce by 15 weeks.[2] At that time, the "aged" calli were subdivided and placed on regeneration medium (the same basic formulation with 2,4-D decreased to 0.25 mg 2,4-D per liter) in petri dishes. The dishes were returned to continuous light (15 μmol m^{-2} s^{-1}) at 23 ± 2°C for the regeneration period. Shoots regenerated from the various callus pieces in about a month. When the shoots were about 1 cm tall, they were transferred to rooting medium which contained no 2,4-D and decreased levels of sucrose and agar, as discussed in Chapter 4 of this book.

The mass of callus per initial explant was quite small because the procedure (Figure 1) allowed only one subculture following callus establishment. Therefore, the number of plants regenerated from each haploid plant was relatively low. Somatic chromosome numbers in the plants that were regenerated from "aged" callus from cytologically verified (21-chromosome) haploid tall fescue plants are shown in Table 1.

TABLE 1
Somatic Chromosome Numbers in Plants Regenerated from "Aged" Somatic-Derived Callus Tissue from Cytologically Verified (21-Chromosome) Haploid Plants of Tall Fescue

Haploid source plant	No. of regenerated plants	No. of regenerated plants with indicated somatic chromosome numbers					
		21	38	40	41	42	81—84
No. 6	7	6				1	
No. 11	6	5				1	
No. 14	7	5		1		1	
No. 15	19	8		2	1	8	
No. 18	4	2	1			1	
No. 20	17	11				2	4
Total	60	37	1	3	1	14	4

From Kasperbauer, M. J. and Eizenga, G. C., *Crop Sci.*, 25, 1091, 1985.

III. EVALUATION

After regenerating plantlets from the "aged" callus tissue, they were evaluated in several ways to assess potential value in the tall fescue improvement program. In progressive order, the evaluations consisted of visual examination of the newly regenerated plantlets; cytological examination of shoot and root tips; exposure to various stress conditions in controlled environments; and field evaluation for plant form, winter survival, and fertility.

A. Visual

The regenerated plantlets were visually examined at frequent intervals, beginning even before they were transferred from regeneration medium to rooting conditions. The earliest visual examination allowed early identification of green vs. albino plantlets. This allowed determination of the frequency of albinos or other obviously abnormal plant types that may have been associated with various culture conditions and genotypes. The procedure outlined in Figure 1 resulted in very few albino or other abnormal-appearing regenerated shoots.[2]

Visual examinations were continued as the plants progressed through the various stages of regeneration, rooting, early growth, and field evaluation. A few plants developed good leaves but poor root systems, causing them to deteriorate rapidly after transfer to soil. A few other plants had visually detectable abnormalities and were discarded before further time-consuming evaluations. The important point is that tissue selection, culture media, physical environment, and timing of the various steps resulted in many green, normal-appearing plantlets that rooted easily and grew well after transfer to soil.

FIGURE 2. An angrogenetic haploid (left) derived via anther-panicle culture and a doubled haploid derived via "aged" somatic tissue culture from the haploid (right). (From Kasperbauer, M. J. and Eizenga, G. C., *Crop Sci.*, 25, 1091, 1985.)

An androgenetic haploid (21-chromosome) tall fescue plant and a doubled haploid (42-chromosome) plant derived via "aged" somatic tissue culture from the haploid are shown in Figure 2. The plants shown in the photograph were each started from a single tiller on the same date and grown under the same controlled environment prior to being photographed. Note that both plants have the same leaf characteristics even though the doubled haploid is larger. This is related to cell size,[11] and the relative size difference between haploid and doubled haploid tall fescue plants is consistent with differences previously observed between tobacco haploids and doubled haploids.[6]

B. Cytological

The regenerated plants that appeared normal and grew well were examined cytologically to determine whether the tissue culture approach was successful in regenerating doubled haploid plants from the selected cytologically verified haploid plants. There were relatively few regenerated plants because the procedures used to initiate the callus, "age" the callus, and regenerate plantlets were designed to minimize the probability of inducing the mutations that occur with numerous subcultures.

At least one cytologically verified 42-chromosome plant was obtained from each of the 6 selected haploid plants used in this tissue culture procedure (Table 1). The doubled haploid (42-chromosome) plants had the same markers (leaf angle, shape, etc.) as the "parent" haploid plant. Some of the haploids,

FIGURE 3. Regenerated 21-, 42-, and 84-chromosome plants (left to right) that were derived via somatic tissue cultures from the same cytologically verified 21-chromosome plant (No. 20, see Table 1).

all of the doubled haploids, and one 84-chromosome regenerated plant from this series were cloned via tillering and taken through further evaluations and comparisons.[2,11,12]

C. Controlled Environment

After cloning, the 21-, 42-, and 84-chromosome plants that were derived from "parent haploid plant no. 20" offered the opportunity to study ploidy effects on various plant characteristics. Representative plants of the three ploidies are shown in Figure 3. All three were started from single tillers on the same date and grown in the same environment prior to being photographed. Leaf size and stem diameter were progressively larger with increased ploidy. These three ploidy levels offered unique material in which to study the influence of ploidy on certain forage-related characteristics (Table 2). Increased ploidy resulted in higher succulence, crude protein, *in vitro* dry matter disappearance, potassium, calcium, and magnesium.[11] However, the 84-chromosome plants were less winterhardy. Also, the potassium concentration resulted in a higher potassium/(calcium + magnesium) equivalent ratio, which has been associated with grass tetany,[11] a physiological disorder of grazing cattle. These preliminary comparisons of ploidy effects on forage characteristics and winterhardiness offer some insight into potential advantages and disadvantages of developing or introducing higher ploidy forage grasses.

TABLE 2
Leaf *In-Vitro* Dry Matter Disappearance (IVDMD), Crude Protein, K, Ca, Mg Concentrations[a] as Influenced by Ploidy Level in Tall Fescue

	Somatic chromosome no.		
	21	42	84
IVDMD	695 ± 6	715 ± 7	729 ± 5
Crude protein	219 ± 4	227 ± 3	239 ± 8
K	32.2 ± 0.6	34.7 ± 0.9	41.6 ± 0.7
Ca	6.8 ± 0.1	6.6 ± 0.5	7.4 ± 0.2
Mg	3.3 ± 0.1	3.3 ± 0.1	3.6 ± 0.1

[a] Values are means ± SE and are given as grams per kilogram of dry matter.

From Kasperbauer, M. J., Karlen, D. L., and Burton, H. R., *Crop Sci.*, 27, 1081, 1987.

D. Field

Even though the visual and cytological evaluations indicated that 42-chromosome normal-appearing plants could be cultured from cytologically verified 21-chromosome haploid plants, the final test was that of fertility under field conditions. To test the fertility of the doubled haploid plants, some were grown in field plots surrounded by seed-derived "Kentucky 31" tall fescue plants. Seeds were collected from a number of the doubled haploid lines and from some of the control (Kentucky 31) plants. Randomly selected seeds from such collections were tested for germination on water-moistened filter paper in petri dishes. The seed collected from the doubled haploid plants germinated in percentages as high or higher than seed collected from control plants. When the radicles were about 1 cm long, the seedlings were transferred to individual pots of potting soil in a growth chamber along with control plants. The seedlings from the tissue culture-regenerated doubled haploids grew rapidly and tillered normally in the growth chamber. At this time (1989), many seeds that were collected from the regenerated doubled haploid plants are in cold storage awaiting field trials to compare the progeny of the doubled haploids with conventionally derived plants. The evaluations (both cytological and agronomic) should provide invaluable information on the feasibility of using these biotechnologically derived materials in conjunction with conventionally derived materials in a tall fescue improvement program.

IV. CONCLUDING REMARKS

Cytologically verified haploid plants[3] of tall fescue were used in several approaches to the development of fertile doubled haploid plants. Colchicine treatments were used as suggested by previous literature.[5] However, none of the attempts with colchicine resulted in any doubled haploid plants.[2]

The second approach was a major revision[2] of a procedure that was previously developed for tobacco.[6] With tobacco, it is possible to regenerate doubled haploid plants directly from excised midveins of fully expanded (''aged'') leaves of a cytologically verified haploid plant.[6] When such haploid tobacco tissue is placed on a culture medium with a tenfold increase in kinetin, some endomitotic divisions occur and are followed by regeneration of some plants with the doubled chromosome number.[6] Major adaptations were needed before success could be attained with tall fescue. Two major differences were that fully expanded leaf midveins of tall fescue would not form callus or regenerate plants, and young tall fescue tissue did not respond to kinetin in the culture medium.[8]

The successful approach with tall fescue[2] consisted of (1) initiating callus from ''young'' stem tissue from rapidly elongating peduncles of cytologically verified haploid plants, (2) placing the new callus on fresh maintenance medium for one prolonged (15-week) subculture to ''age'' it (analogous to the ''aging'' tobacco leaf[6]), and (3) placing the ''aged'' callus tissue (in which the rate of cell division had slowed considerably) on plant regeneration medium. Since tall fescue callus appears to be autonomous for cytokinin, the decreased auxin/cytokinin ratio was accomplished by decreasing auxin content in the tissue culture medium.

The plants that were regenerated from cytologically verified (21-chromosome) haploids of tall fescue included some with 21, some with 42, and a few with 84 chromosomes. The three ploidies from a given haploid ''parent'' plant were useful in evaluation of ploidy effects on various forage characteristics. However, the major findings were that (1) the 42-chromosome (doubled haploid) plants had the same plant ''markers'' as the 21-chromosome ''parent'' haploid plants, and (2) that the doubled haploid plants were fertile and produced abundant viable seed when grown in field plots.

REFERENCES

1. **Hansen, A. A.,** The future of tall fescue, in *Tall Fescue Monograph,* Buckner, R. C. and Bush, L. P., Eds., American Society of Agronomy, Madison, WI, 1979, 341.
2. **Kasperbauer, M. J. and Eizenga, G. C.,** Tall fescue doubled haploids via tissue culture and plant regeneration, *Crop Sci.,* 25, 1091, 1985.
3. **Kasperbauer, M. J., Buckner, R. C., and Springer, W. D.,** Haploid plants by anther-panicle culture of tall fescue, *Crop Sci.,* 20, 103, 1980.
4. **Eizenga, G. C. and Kasperbauer, M. J.,** Chromosome pairing in tall fescue haploids derived by anther-panicle culture, *J. Hered.,* 76, 99, 1985.
5. **Buckner, R. C., Hill, H. D., and Burrus, P. B.,** Some characteristics of perennial and annual ryegrass × tall fescue hybrids and of the amphidiploid progenies of annual ryegrass × tall fescue, *Crop Sci.,* 1, 75, 1961.

6. **Kasperbauer, M. J. and Collins, G. B.,** Reconstitution of diploids from leaf tissue of anther-derived haploids in tobacco, *Crop Sci.*, 12, 98, 1972.
7. **Jensen, C. J.,** Chromosome doubling techniques in haploids, in *Haploids in Higher Plants: Advances and Potential,* Kasha, K. J., Ed., University of Guelph Press, Guelph, Ontario, 1974, 153.
8. **Kasperbauer, M. J., Buckner, R. C., and Bush, L. P.,** Tissue culture of annual ryegrass × tall fescue hybrids: callus establishment and plant regeneration, *Crop Sci.*, 19, 457, 1979.
9. **Clapham, D.,** Haploid *Hordeum* plants from anthers in vitro, *Z. Pflanzenzuecht.*, 69, 142, 1973.
10. **Wilson, H. M.,** Culture of whole barley spikes stimulated high frequencies of pollen calluses in individual anthers, *Plant Sci. Lett.*, 9, 233, 1977.
11. **Kasperbauer, M. J., Karlen, D. L., and Burton, H. R.,** Ploidy effects on protein in vitro dry matter disappearance, and potassium/(calcium + magnesium) equivalent ratio in tall fescue forage, *Crop Sci.*, 27, 1081, 1987.
12. **Kasperbauer, M. J.,** Tall fescue plant modification through tissue culture, haploids and doubled haploids in *Proc. 6th Int. Congr. of Plant Tissue and Cell Culture,* University of Minnesota Press, Minneapolis, 1986, 124.

Chapter 7

CYTOLOGY OF HAPLOIDS AND DOUBLED HAPLOIDS

Warren D. Springer

TABLE OF CONTENTS

I. INTRODUCTION

The point is often made, when dealing with a polyploid species, that increased understanding concerning the genetic structure of the species would be beneficial for useful manipulation, i.e., breeding, of the species. Several different aspects of "genetic structure" are implied when such suggestions are put forth. First, we want to ascertain whether the species is an autopolyploid or an allopolyploid. Secondly, we would like to determine the degree of relatedness between the component genomes, whether auto- or allopolyploid. Additionally, we would like to be able to determine the progenitor species that have contributed to the genetic structure, in the case of allopolyploids. This is of importance in the determination of the size of the gene pool available for potential exploitation when attempting to improve the species.

Ploidy level is a simple concept on the most fundamental level. Ploidy, as defined in *A Glossary of Genetics and Cytogenetics,*[1] is "a term referring to the number of chromosome sets per cell." Further, "polyploidy" is defined, in the same reference, as "somatic cells and tissues, as well as (of) individuals having three, four, five, or more complete chromosome sets instead of two as in diploids." deWet[2] uses the definition "polyploids combine three or more basic genomes of the taxonomic group to which they belong." What constitutes a "complete chromosome set" or a "basic genome" may not always be apparent. However, this can result in confusion of terminology and/or the development of new terminology to specify particular situations. The use of the term "haploid", when dealing with polyploids, has been a prime example of the confusion which can occur. This is clearly attested to by the inclusion of a chapter on terminology in the Proceedings of the First International Symposium on Haploids in Higher Plants.[3]

In this discussion, the term haploid will be used to refer to any cell, tissue, and/or individual having the normal gametic chromosome number. This designation will be used consistently regardless of the ploidy level of the species.

In the absence of haploids, the behavior of chromosomes in interspecific and intergeneric crosses provided the information from which assessment of genome relatedness is made. This is very useful information and has been crucial to our understanding of a number of polyploid species. However, a critical component in developing a full understanding of the nature of the genome structure of a polyploid is the careful study of the cytogenetic behavior of haploids of the species. Haploids permit us to examine chromosomal relationships in the absence of any genetic chromosomal "pairing mechanisms" inherent in the second parent in interspecific or intergeneric crosses. The utility of haploids in understanding chromosomal relatedness is well documented in a number of species.[4-6] The intent of this chapter is to review the current level of our understanding of the cytogenetics of haploid tall fescue and to suggest where additional information would be of benefit.

The genus *Festuca* is cytologically diverse, containing species having chromosome numbers ranging from 14 to 70. The base chromosome number in the species is 7. Malik and Thomas have presented an extensive review of the various species.[7] Tall fescue, *Festuca arundinacea* Shreb., is a hexaploid species having a 2x(= 6n) chromosome number of 42. It has been alternatively suggested in the literature that tall fescue is an allohexaploid[8,9] and that it is an auto-allohexaploid.[7,10] Most recent studies indicate that the genomic structure of tall fescue is that of an allohexaploid, having a genome structure designated by the genomic formula AABBCC.

II. METHODS

A. Tissue Sources and Staining Procedures

There is nothing intrinsic about haploids or doubled haploids that makes them unique in regard to the application of standard cytological techniques. Several different cytological techniques have been utilized in published reports concerning tall fescue haploids and doubled haploids.

The first objective when working with haploids and their derivatives is to accurately establish the chromosome number of each plant included in the study. The primary somatic tissues used to obtain mitotic metaphase spreads are root tips and shoot tips (leaf meristems). The root tip procedure given by Morgan[11] was used by Kasperbauer et al.[12] in the first report of the production of *in vitro*-derived tall fescue haploids. Morgan's procedure is a very simple process of removing 1- to 2-cm-long root tips and pretreating them in 1°C distilled water for a period of 16 to 17 h. The root tips are then fixed in 3:1 absolute ethanol:glacial acetic acid for a minimum of 2 h. The root tips are prepared for cytology by hydrolysis in 1 *N* HCl at 60°C for 10 min. The root tips are stained in Feulgen stain at room temperature. Following the development of sufficient stain intensity, normally requiring at least 30 min, the root cap is removed and the meristematic region squashed in a drop of acetocarmine on a slide.

An alternative procedure was utilized by Kasperbauer and Eizenga.[13] The root tips were pretreated in 4°C distilled water for approximately 20 h. The pretreatment was continued in a saturated solution of mono-bromonaphthalene at room temperature for 3 h. After rinsing in distilled water, the root tips were fixed in 3:1 95% ethanol:glacial acetic acid for a period of 30 min. After rinsing in distilled water again, the root tips were stained in propionocarmine for 12 h. The root tips were then hydrolysized in 6 *N* HCl for 10 min at room temperature and then stored in distilled water at 4°C. Squashes were made in propionocarmine for observations.

These procedures simply represent those chosen by the individuals performing the studies. Minor modifications of a number of related procedures would undoubtedly result in satisfactory preparations. As examples, when utilizing the former procedure, the author's experience is that 3:1 fixation of no more than 24 h followed by transfer to 70% ethanol for a minimum of several days improves the quality of the preparations. Additionally, transfer from the 70% ethanol to distilled water for a period of several minutes prior to hydrolysis is helpful.

An alternative to the use of root tip tissue is the use of shoot tips. The use of shoot tips was preferred by Kasperbauer et al.[12] because a higher mitotic index was observed in shoot tips compared to root tips. Additionally, cell disruption was less frequent when using shoot tips. Figure 1 shows an example of the preparations obtained from the use of leaf meristems according to the procedure used by Kasperbauer et al.[12]

The shoot tip procedure is a modification of the technique developed by de Lautour,[14] as modified by Bennett.[15] The procedure begins with the excision of young actively growing tillers from either greenhouse- or field-grown plants. The meristems are pretreated for 3 h in a saturated solution of a-bromonaphthalene in 0.05% saponin in water at room temperature. The tissue is fixed in a 3:1 ethanol:glacial acetic acid solution and stored in 70% ethanol. The fixed material is prepared for cytological observation by maceration in an enzymatic solution of 5% pectinase and 2% cellulase in 0.1 *M* acetate buffer at pH 3.6 for 3 h. The tissue is stained in 2% orcein, and squashes made by transferring a small piece of tissue to a drop of propionocarmine stain on a slide. The tissue is then macerated with the end of a glass rod prior to squashing under a cover glass on a slide.

Meiotic tissue can also be used for the determination of chromosome numbers in haploids and doubled haploids. Pollen mother cells have been utilized by several different investigators for the determination of chromosome numbers. The standard procedure is to fix panicles in Carnoy's solution A (6:3:1 ethanol:acetic acid:chloroform) containing a small amount of ferric chloride. Anthers are removed and opened in a drop of propionocarmine, and the pollen mother cells teased out for staining and subsequent slide preparation.

Although pollen mother cells are the normal objects of attention when utilizing these tissues, the author has had very good success utilizing tapetal cells in another polyploid grass species for the determination of somatic chromosome numbers. The metaphase spreads obtained are very good, and good separation of up to 72 chromosomes has been obtained. Another advantage of using tapetal cells is that a very large number of metaphases can often be observed on a single slide due to the syncrony of the final nuclear division among cells comprising this tissue. The final nuclear division of the tapetal cells generally coincides with prophase I. It is not an uncommon experience to find that meiosis becomes irregular in genetically unbalanced situations, i.e., wide crosses and changes in ploidy level. The irregularity is seen as an expanded range of meiotic stages present within individual anthers and a less regular sequential pattern of development along the panicle. This meiotic irregularity does not appear to affect the syncrony of the nuclear division in the tapetal cells.

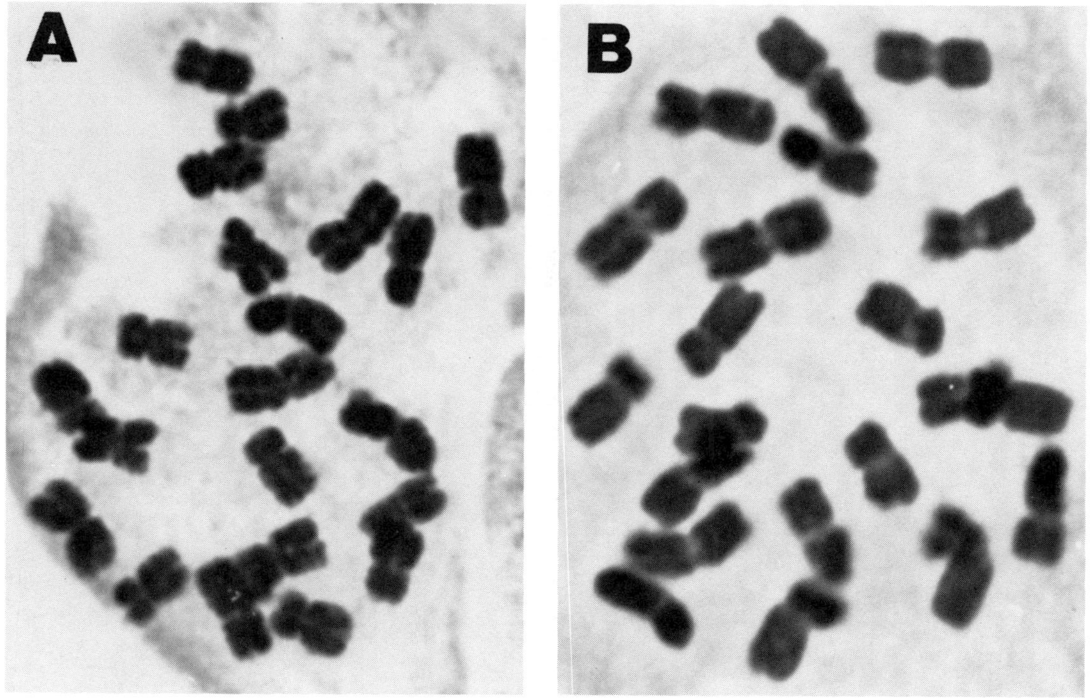

FIGURE 1. Mitotic metaphase preparations of haploid tall fescue showing 21 chromosomes. The tissue utilized for these preparations was leaf meristem.

B. Problems

Haploids present no unique challenges relative to standard mitotic cytology. The chromosomal preparations from root tip and shoot tip material are of equivalent character and quality as those prepared from chromosomally unreduced tissues. However, as has been alluded to above, meiotic cytological observation does involve some difficulties not routinely encountered when studying chromosomally unreduced materials. The difficulties are not unlike those encountered when working with any genetically unbalanced materials.

An immediate problem when attempting meiotic examination of haploid tall fescue is the generally reduced size of the panicle in haploids. Therefore, the amount of material available within each panicle from which to obtain appropriate meiotic stages for examination is greatly reduced. This requires that the haploid plants be carefully observed for the appearance of flowering stalks and that care be taken to collect panicles at the appropriate stage of development in order to maximize the amount of material available to work with. This problem is further compounded by a reduction in the syncrony of meiosis within each anther relative to the syncrony of meiotic stage found in chromosomally unreduced materials. Consequently, it is not unusual to find a broad range of meiotic stages both within and between anthers from the same floret. This can make meiotic cytology a more rigorous effort than that required when working with chromosomally unreduced plant material.

Additionally, the interpretation of observed meiotic figures is often less than routine. A common condition found in meiotic figures is a condition referred to as "stickiness". Stickiness is observed as a tendency for chromosomes to clump together and the appearance of an amorphous material attached to and extending between chromosomes. The material stains similarly to the nucleolus, and some investigators have suggested that it represents remnant nucleolar material. This condition can affect both the general spreading of the chromosomes at different stages of meiosis and the observation and interpretation of individual pairing relationships among chromosomes. No satisfactory explanation of the cause nor procedure to reduce the condition has been suggested. However, the appearance of the condition is not unusual, and it must be contended with in obtaining satisfactory preparations and in their interpretation.

III. OBSERVATIONS

There are a limited number of accounts in the literature reporting the recovery of haploids of tall fescue. The first report was by Rogers[16] in 1964. That report was followed by the report of Malik[17] in 1967. In that report, Malik records the recovery of haploids among the progeny of intraspecific crosses involving *Festuca arundinaceas* of different ploidy levels. The haploids were derived from the crossing of $2n (= 6x) = 42$ *F. arundinacea* $\times$ $2n (= 4x) = 28$ *F. arundinacea*. That report simply records the occurrence of the haploids, and no indication of any cytological analysis is given.

The first report of haploids in tall fescue which involved cytological meiotic examination was by Malik and Tripathi[18] in 1970. Only one haploid plant, obtained as a twin seedling, was available for examination. However, a sufficient number of pollen mother cells were observed and recorded to provide useful information. The results of the examination were that the mean chromosomal pairing frequency at metaphase I was 12.96 univalents (I) + 4.01 bivalents (II). Additionally, fully 75% of the cells examined displayed at least 4 II and 40.6% had at least 5 II. The greatest degree of pairing observed was 7 I + 7 II, and that was seen in 7.8% of the cells recorded.

The next published report of haploids in tall fescue was by Kasperbauer et al.[12] in 1980. Again, as with Malik's earlier report,[17] this report did not include any cytological meiotic analysis of the haploids. It simply records the determination of the chromosome number of a number of *in vitro*-derived plants. This was the first report of the *in vitro* production of haploids, however, and brought with it the means for producing large numbers of haploids for utilization in both basic and applied studies.

The last report strictly concerning haploids was by Eizenga and Kasperbauer[19] in 1985. This report was, as was Malik and Tripathi's,[18] a cytological meiotic examination of haploid material. In this instance there were several haploid plants involved, and they were derived via *in vitro* culture as described by Kasperbauer et al.[12] Eizenga and Kasperbauer observed approximately 50 pollen mother cells in 6 separate haploids. The six plants were divided into two groups based on the mean chromosome pairing frequencies observed. On one hand, three plants had mean chromosomal pairing frequencies of greater than 20 I and less than 0.31 II, whereas the other group had a maximum of 18.1 I and a minimum of 1.33 II. Unpublished observations by the author, using haploid plants from the same source population, are consistant with those reported by Eizenga and Kasperbauer.

The published reports concerning cytological observations in haploids have been confined to reporting observations of somatic mitotic metaphase and meiotic metaphase I. Observations concerning other stages of meiosis are lacking. However, it would be instructive to have a more complete picture of meiosis in haploids.

As in any species, the early prophase stages of meiosis are not particularily instructive from gross observation. However, distinct features are easily seen at pachynema. As would be expected in a haploid, the degree of synapsis at pachynema is quite variable. The most striking feature is the general high degree of pairing which is seen. The majority of pollen mother cells appear to have extensive pairing with some pollen mother cells appearing to have all chromosome segments paired. Figures 2 to 4 show examples of pollen mother cells which demonstrate the variability of synapsis which is seen in the haploids. This is an unexpected observation, considering the chromosomal constitution of the haploids and the normal expectation concerning chromosomal pairing when homologues are absent. If full homoeologous pairing

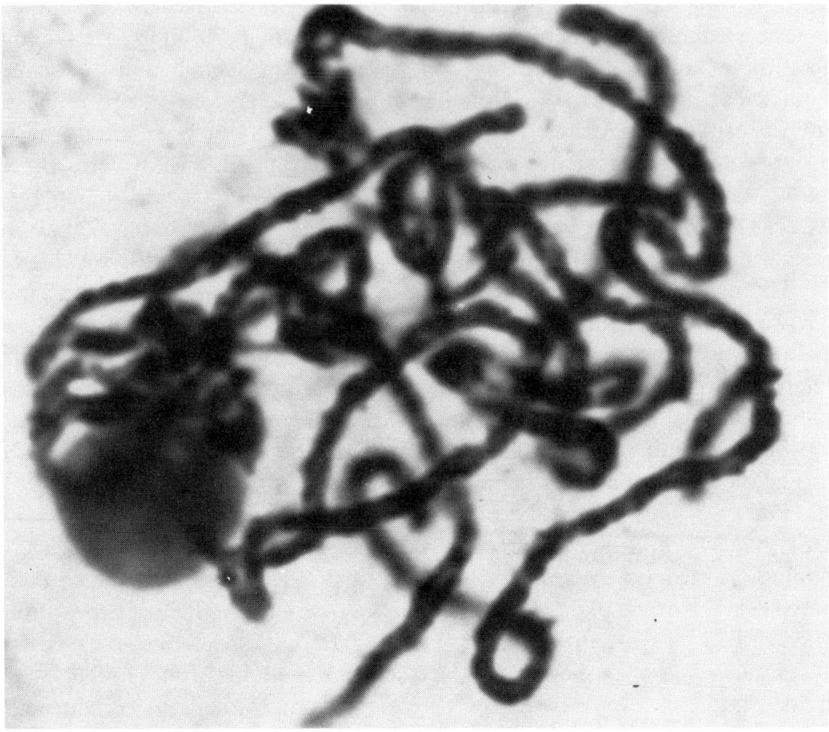

FIGURE 2. Pollen mother cell from a haploid tall fescue plant displaying essentially complete synapsis of all chromosomes present at pachynema.

occurred between two of the sets of seven chromosomes, the remaining set would have no pairing partners and any pairing involving those chromosomes would be illegitimate pairing between nonhomologues. Alternatively, if the three chromosome sets are equally similar (or dissimilar, as the case may be), then the expectation would be to observe trivalent pairing, i.e., chromosomes alternately paired with different homoeologues along their length. Although instances of switching between pairing partners can be found, it is not observed with any appreciable frequency. There appears to be a fair degree of foldback pairing of individual chromosomes within the haploids. In such a case it would be expected that either unpaired loops or unpaired ends would be observed. Both of these situations are obvious in cells having extensive pairing, as shown in Figure 5.

However, it is not possible to obtain sufficient spreading of the pachytene chromosomes for utilizing pachynema for routine chromosomal analysis. It is only on occasion that a pachynema spread is observed in which a complete paired chromosome is sufficiently separated from the remaining chromatin to be studied in any detail. Figures 6 and 7 are indicative of such a situation.

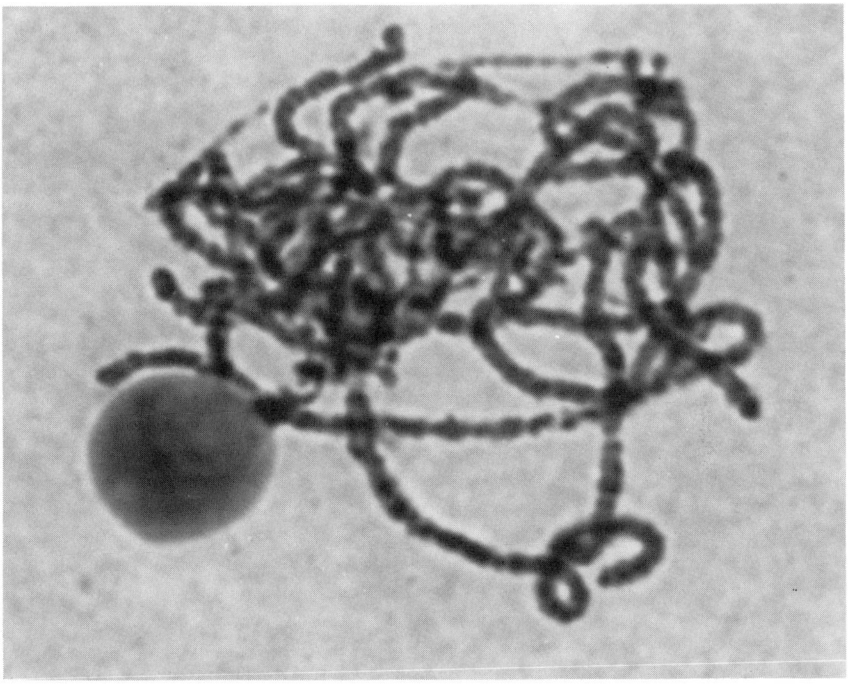

FIGURE 3. Pollen mother cell from a haploid tall fescue plant displaying a combination of apparently well-synapsed regions and unpaired chromosomal regions at pachynema.

It can be seen in Figure 7 that individual chromosomes can be studied in circumstances where single chromosomes are separated from the remaining chromosomal mass. The chromosome depicted in Figure 7 is shown to be a single chromosome that has had the two chromosomal arms fold back on themselves, resulting in the appearance of a ''typical'' pachytene bivalent. The process of screening a sufficient number of pollen mother cells at pachynema and observing a large enough number of separate chromosomes would allow one to identify all the chromosomes in great detail. However, such a task would be extensive, and the value is questionable. The primary point to be made here is that the apparent high degree of bivalent pairing initially observed when first viewing pachynema in haploids can be misleading.

Following pachynema, the chromosomes become individually distinguishable during diplonema and diakinesis (Figure 8). The actual degree of chromosome-to-chromosome association begins to emerge at this stage of meiosis. It is apparent from observation of diplonema/diakinesis that much of the pairing seen at pachynema is merely loose association of chromosomal regions. Typical observations are shown in Figures 9 to 12.

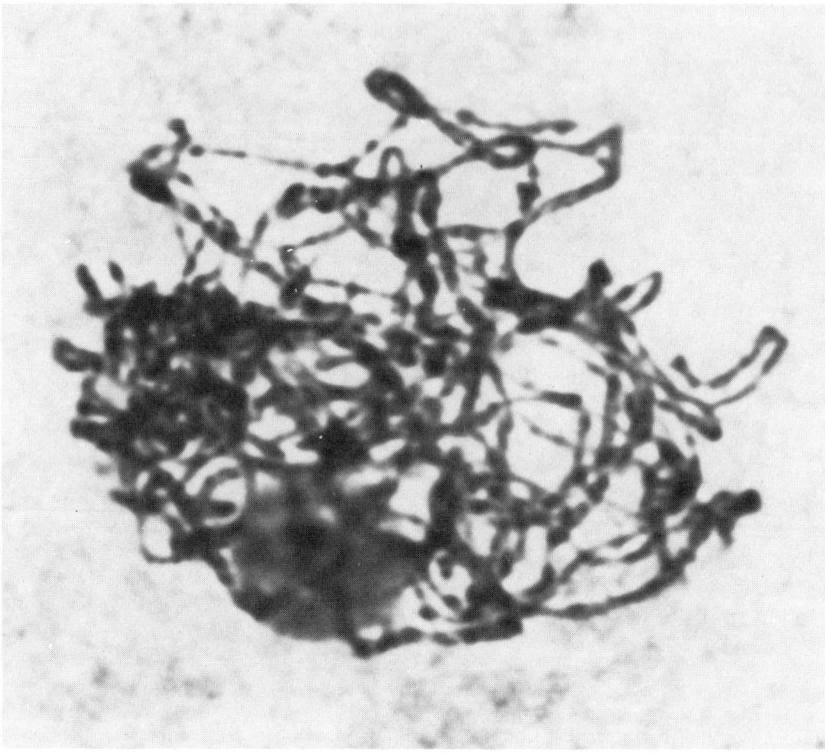

FIGURE 4. Pollen mother cell of haploid tall fescue at pachynema displaying almost complete asynapsis for all chromosomes.

Haploid metaphase is certainly the most instructive stage of meiosis relative to interpretation of chromosome relatedness between component chromosome sets. Metaphase I in normal hexaploid tall fescue is as would be expected when strict autosyndetic pairing is the rule in a hexaploid. Typical rod- and ring-type bivalents are routinely seen, as shown in Figure 9. These same structures can be seen in metaphase I spreads of haploids of tall fescue (Figures 10 to 12). Additionally, some multivalent structures are observed in some haploid metaphase I spreads, as seen in Figure 11. However, as indicated above, the problem of ''stickiness'' can make the analysis difficult. As shown in Figure 13, a typical metaphase I spread is not as clearly interpretable as one would like. Figure 13 clearly shows 21 chromosomes. While the chromosomes arbitrarily numbered 14 and 15 appear to form a typical rod bivalent, it is a matter of interpretation as to what degree of real association exists between the chromosomes which are included in the groupings containing the chromosomes numbered 4 to 7, 8 to 11, and 17 to 21. Figure 14 shows another typical metaphase I. In this metaphase, no chromosomes appear to represent

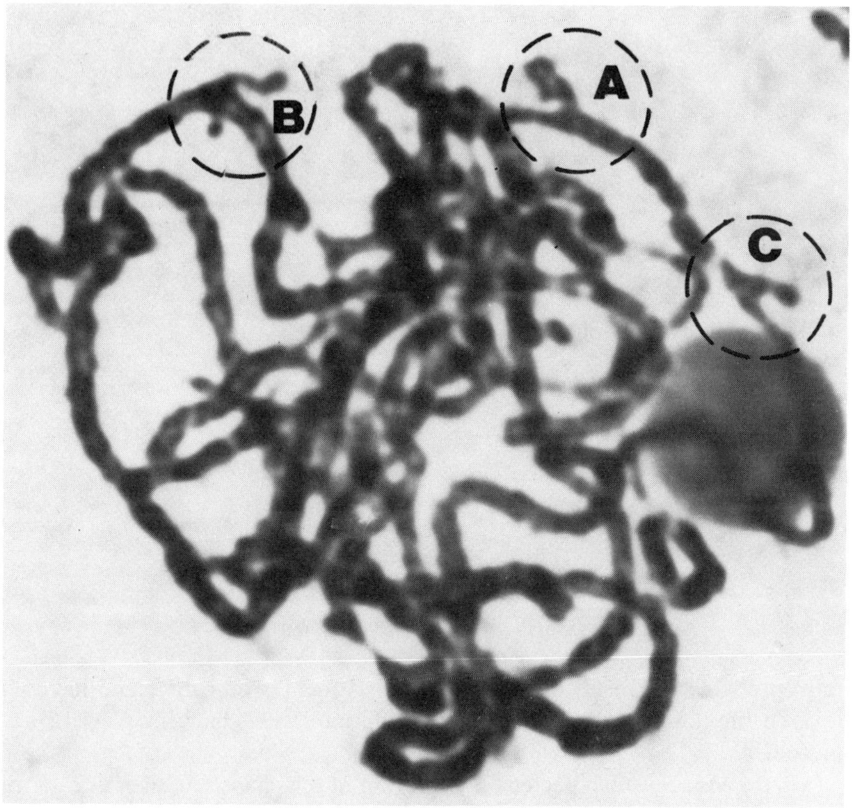

FIGURE 5. Pollen mother cell of haploid tall fescue displaying fold-back pairing (A), and unpaired, mismatched ends (B) and (C).

typical rod bivalents. The majority of chromosomes appear as univalents. However, the chromosome group numbered 7 to 13 is less definitive in interpretation. These problems result in the necessary observation of a larger number of metaphase cells than would be required in some other materials in order to obtain a sufficient number of unambiguous spreads to provide useful information.

The stages of meiosis following metaphase I become increasingly more difficult to utilize due to the breakdown of meiosis caused by the unequal distribution of chromosomes at anaphase I. Several examples of the unequal distributions which are typical are shown in Figures 15 to 17. It is during this phase that evidence for pairing and recombination between structurally dissimilar chromosomes could be seen by the occurrence of chromatin bridges and fragments. Bridges are observed at anaphase I, as seen in Figures 18 and 19, however, no chromatin fragments are seen. The bridges observed are

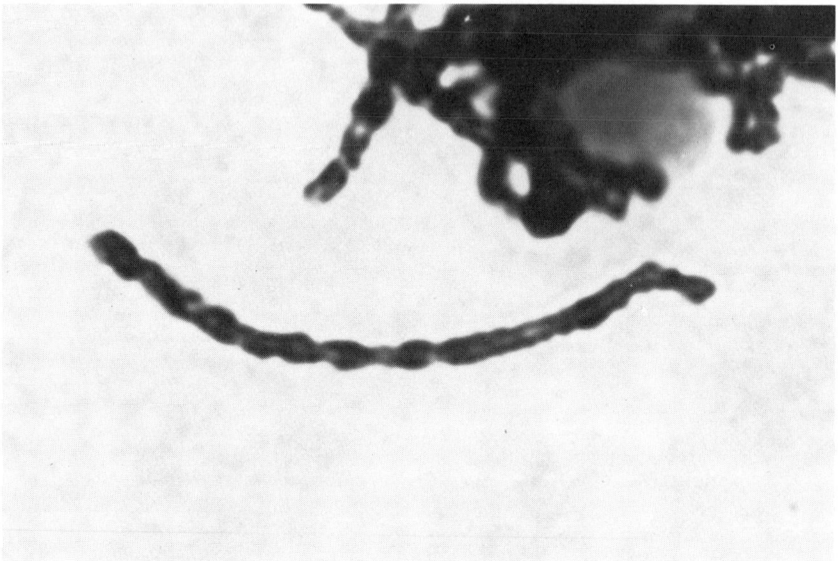

FIGURE 6. An isolated chromosome at pachynema.

most probably the result of the "stickiness" mentioned earlier and have no relationship to interpretation of structural dissimilarities between paired chromosomes.

A second, not unexpected, observation at anaphase I is the appearance of laggards. The occurrence of laggards is typical in situations where univalents are present during the first meiotic division. Figure 20 displays a typical anaphase I cell demonstrating the presence of a laggard. While some of these lagging chromosomes may migrate to the poles belatedly and be included in the respective interphase nuclei, some end up as micronuclei and are excluded from the interphase nuclei (Figure 21). Additionally, it is not uncommon to observe precocious division of univalent chromosomes at anaphase I. This can be seen both at anaphase I (Figure 22) and at prophase II (Figure 23). The observation of single chromatids at interphase is evidence of the precocious division of a univalent at the previous anaphase.

The preceding discussion clearly indicates the extent of meiotic disruption in the haploids of tall fescue and gives an indication of what one should expect to observe when undertaking cytological meiotic examination of similar plants.

Returning to the data of Malik and Tripathi[18] and Eizenga and Kasperbauer[19] collected by observation of metaphase I, it is evident that the results reported from those two studies do not correspond well. Eizenga and Kasperbauer observed mean chromosomal pairing frequencies of from 20.66 I + 0.17 II to 15.98 I + 2.36 II + 0.10 III for individual plants examined. None of

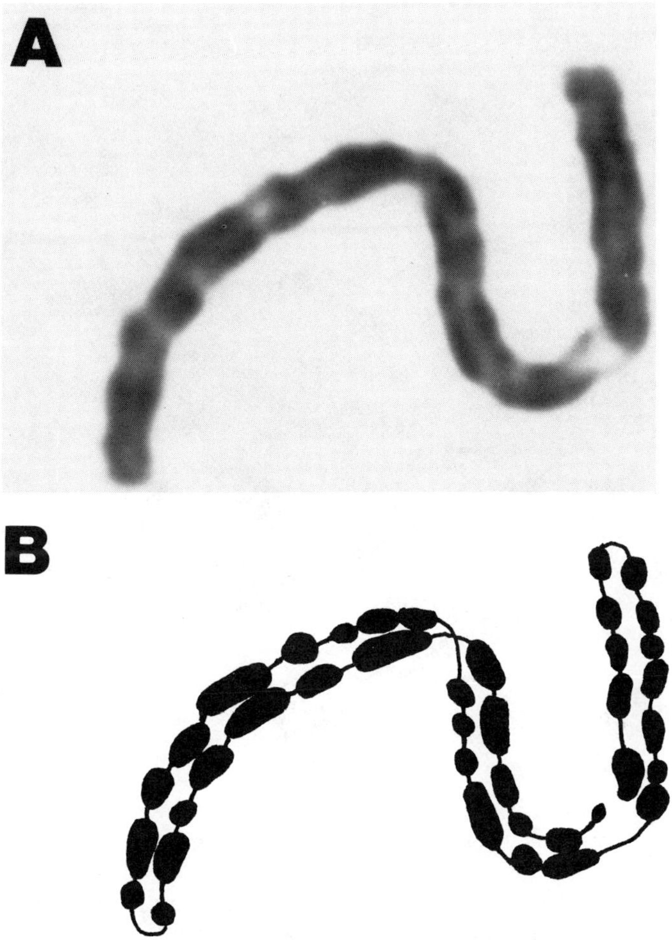

FIGURE 7. A single pachytene chromosome displaying fold-back pairing. A photomicrograph (A) and a diagrammatic representation (B) of the major chromomeres developed from examination of a number of photomicrographs produced using varying exposure and development times.

these frequencies approximate the 12.96 I + 4.01 II observed by Malik and Tripathi in the single-haploid plant utilized in their study. The most obvious explanation for this discrepancy is the genetic background of the source material. Tall fescue is a genetically diverse species. Using seed protein electrophoresis, Bulinska-Radomska and Lester[20,21] have very adequately demonstrated the diversity among hexaploid *F. arundinacea*. The range of Jaccard's Similarity Coefficients (Sj) between various races of tall fescue was so broad that it overlaps the range of Sjs arrived at when comparing tall fescue to *F. gigantea*. Additionally, chromosome structural heterogeneity has been demonstrated between different populations of tall fescue by Malik.[22]

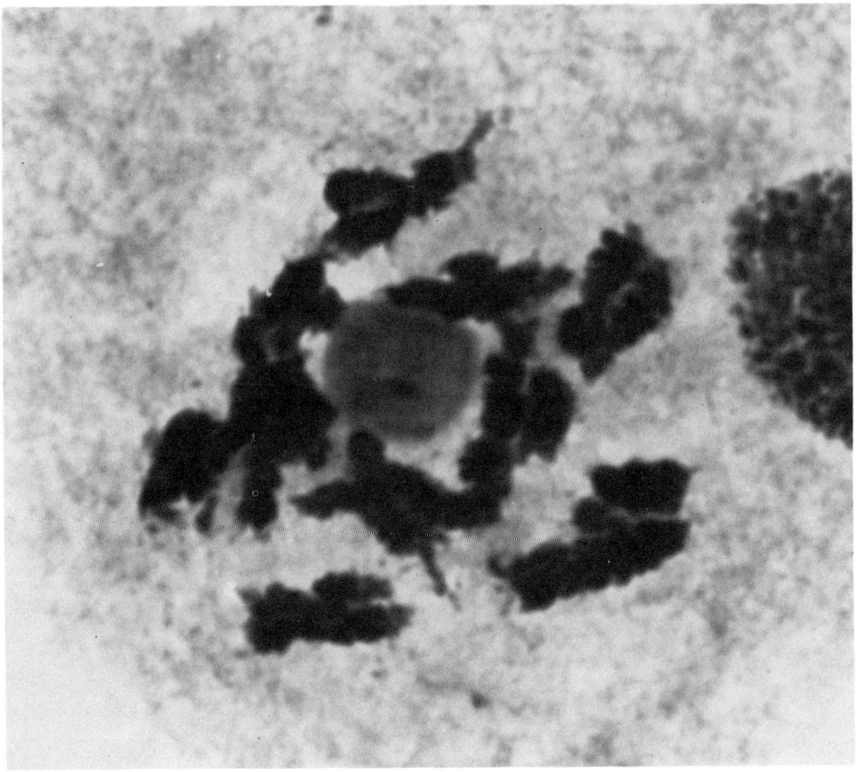

FIGURE 8. Typical diplonema/diakinesis of haploid tall fescue.

The genotypic background of the material used by Malik and Tripathi is not identified in their report. However, it was probably a European or Mediterranean type. Eizenga and Kasperbauer derived their haploids from the North American commercial cultivar known as Kentucky 31. Knowing what we do concerning the inherent diversity in the species, some disparity in results might be expected.

An obvious solution to this problem would appear to lie in the use of *in vitro* techniques to control the diversity. If one were to begin with the haploids derived from anther-panicle culture and then recover dihaploids from them, the resulting plants could again be anther-panicle cultured to generate a second round of haploids. All haploids from a single dihaploid would be expected to be similar and to display a consistent mean chromosomal pairing frequency, as opposed to the results obtained by Eizenga and Kasperbauer. However, there is no reason to expect that all sets of haploids (each derived from a separate dihaploid) would display the same pairing frequencies. The dihaploids would also be very useful in studies of pairing in hybrids (intra- and interspecific and intergeneric).

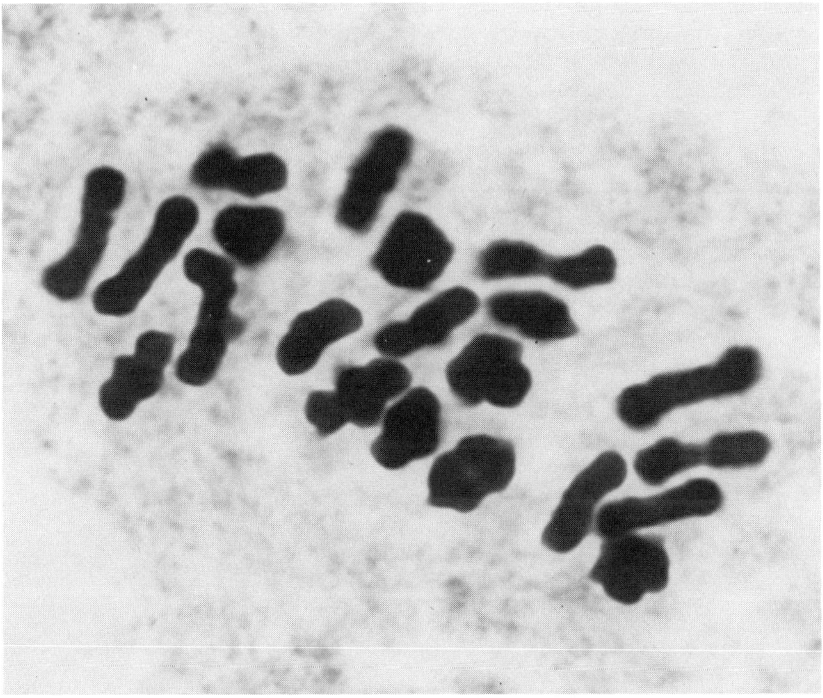

FIGURE 9. Metaphase I spread of normal hexaploid tall fescue (variety "Kentucky 31") showing typical rod and ring bivalents; 21 bivalents are clearly evident.

This points out potential problems in making broad inferences of genetic relatedness when using a very small genetic sample or a source population containing significant variability. The utilization of procedures to control the variation is critical.

IV. APPLICATIONS

A. Phylogenetics

Malik and Thomas published papers in 1966[23] and 1967[7] in which they present evidence for the classification of the genome structure of tall fescue as that of AABBB′B′. The genomic designation of AABBB′B′ has been variously referred to as allo-autopolyploidy[23] and segmental polyploidy.[18] This proposed genomic structure was based on examination of the karyotypes of a number of *Lolium* and *Festuca* species and on the observations of chromosome associations at meiosis in F_1 hybrids between and within species and genera. Gross similarity between one of the tall fescue chromosome subsets and that of *F. pratensis* ($2n = 14$) was also reported. Their observations also

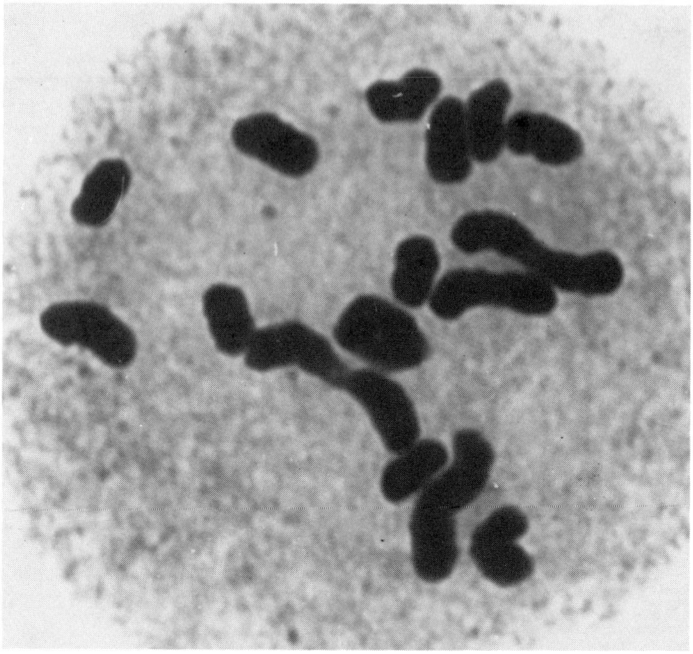

FIGURE 10. Metaphase I spread of haploid tall fescue displaying 4 typical rod bivalents, 1 ring bivalent, and 11 unpaired chromosomes.

indicated that the karyotype of *L. multiflorum* was different from that of *F. pratensis* or any of the chromosome subsets of *F. arundinacea*.

However, Jauhar[8,9] states that tall fescue is an allopolyploid having a genomic designation of AABBCC. Jauhar further states that the extensive pairing observed in hybrids (interspecific and intergeneric) is simply indicative of the homoeology between the component subsets of chromosomes that make up the genome of tall fescue. The meaning of such a statement is not totally clear in that the term "homoeology" is used to describe partially homologous chromosomes. "Homoeology" designates the residual homology of originally completely homologous chromosomes.[1] Therefore, the utilization of the term here implies a phylogenetic relatedness between component chromosome subsets and would, in fact, support the suggestion by Malik and Thomas that the proper genome designation should be AABBB′B′.

Eizenga and Kasperbauer[19] applied the Alonso-Kimber[24] model for determining relatedness of component chromosome subsets to their data and concluded that the suggestion of a genomic designation of AABBCC is the correct one.

The fact that tall fescue is a hexaploid is without question. Nevertheless, the nature of that hexaploidy is an area in which the careful examination of haploids, where variation is controlled, may provide useful information.

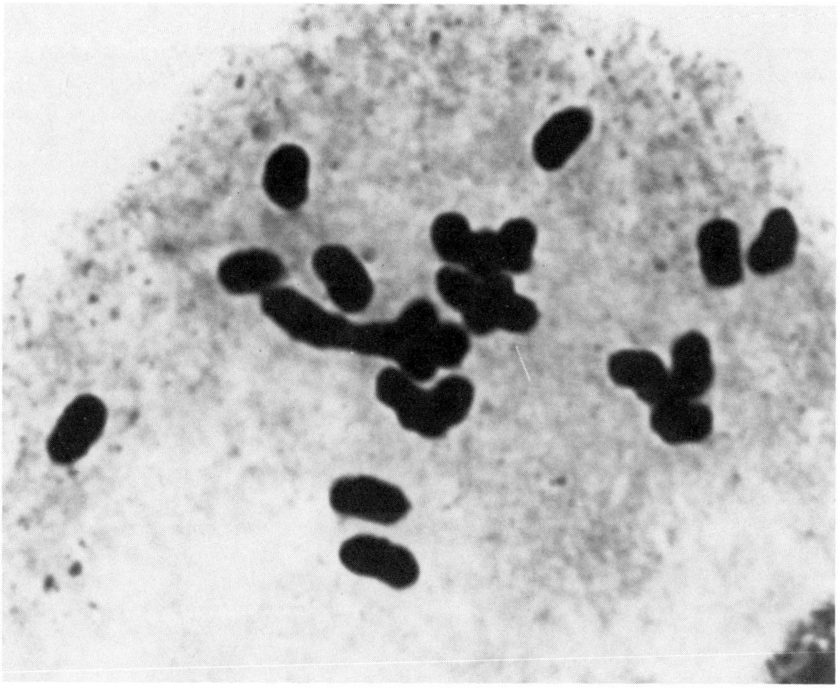

FIGURE 11. Metaphase I spread of haploid tall fescue displaying 3 rod bivalents, 1 apparent trivalent, and 12 unpaired chromosomes.

B. Basic Studies

The utility of haploids for basic studies in genetics and cytogenetics is widely recognized, as attested by the coverage accorded this topic at the First International Symposium on Haploids in Higher Plants[25] in 1974. The most obvious aspect of utilizing haploids in basic studies is that of mutant selection. Selection at the haploid level should permit the observation and potential recovery of all phenotypically expressed mutations. Dominance and recessiveness have no meaning in a true haploid system, since they are relative terms by definition. However, in the case of a polyploid species such as tall fescue, the observation and recovery of mutations may not be as straightforward due to the possible replication of loci between the component chromosome sets which make up the genome. Haploids of polyploids (polyhaploids) are amenable to haploid selection to the degree to which they selectively have become dependent on a single gene system for each metabolic function. This suggests an interesting utility for haploids of tall fescue. A comparative study of the recovery of similar mutations in haploid tall fescue vs. a diploid (2x) species should provide information on the processes of evolution and speciation in polyploids. The preponderance of evidence is that *F. pratensis* is the

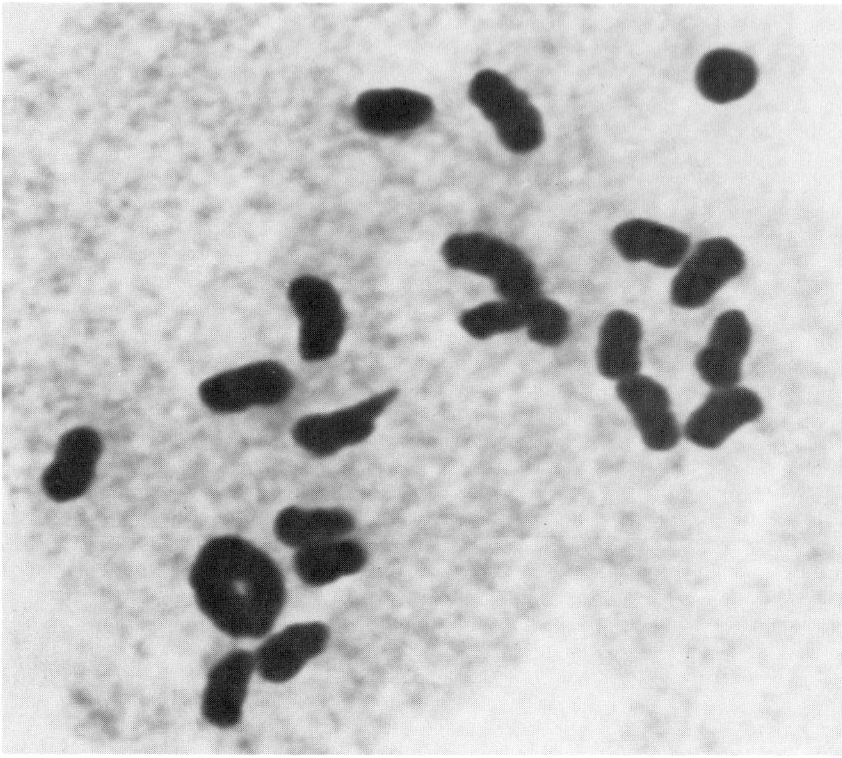

FIGURE 12. Metaphase I spread of haploid tall fescue displaying 1 ring bivalent and 19 unpaired chromosomes.

contributor of the "A" genome in tall fescue. If the *in vitro* anther-panicle culture procedure for the recovery of haploids can be extended to *F. pratensis,* one could compare the mutation rate between haploid (n = x) *F. pratensis,* diploid (2n = 2x) *F. pratensis,* haploid (n = 3x) tall fescue, and diploid (2n = 6x) tall fescue. One might reasonably expect that *F. pratensis* haploid cells or tissue would display more observable mutants than comparable tall fescue material. The rates and ranges of mutants recovered would be interesting.

A second area of basic genetics that can benefit from the careful use of haploids is the study of the genetic control of chromosome pairing in tall fescue. In addition to the information that can be gathered concerning the genetic system that regulates the "diploidization" process in polyploid species, the information may be useful in designing introgression programs for tall fescue breeding.

Jauhar[8,9] was the first to address the genetic control of chromosome pairing in the species. Based on the observation that a particular monosomic displayed

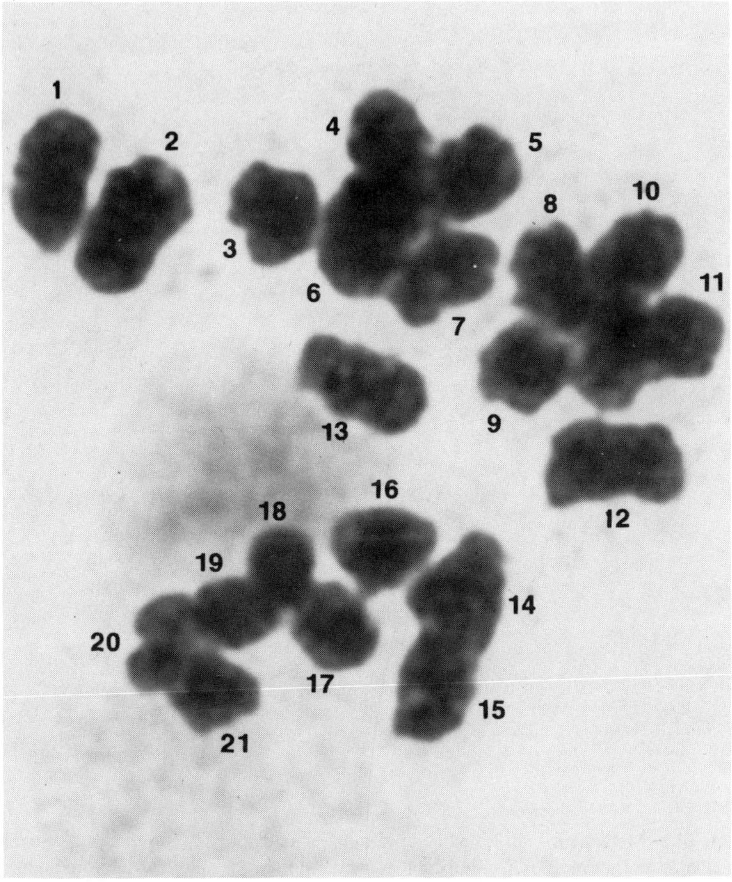

FIGURE 13. Metaphase I spread of haploid tall fescue displaying "stickiness", or clumping, of chromosomes, making interpretation of pairing relationships difficult; 1 "typical" rod bivalent is seen together with 12 unpaired (univalent) chromosomes and 2 groups of ambiguous pairing relationship.

a significantly higher degree of homoeologous pairing than did several other monosomics, Jauhar suggested a hemizygous ineffective genetic pairing control mechanism. The monosomics were derived from a cross of two different tall fescue ecotypes. The parental ecotypes displayed normal "diploid" (bivalent) pairing at meiosis. Euploid progeny from the cross also demonstrated complete bivalent pairing. Several monosomic sibs from the same cross also displayed very high frequencies of bivalent formation with very low multivalent formation. However, as stated above, one particular monosomic had a significantly increased frequency of multivalent formation. Jauhar also cites the occurrence of multivalents in *Lolium* × *Festuca* F_1 hybrids, while the amphiploid derivatives of the F_1s display almost complete bivalent formation

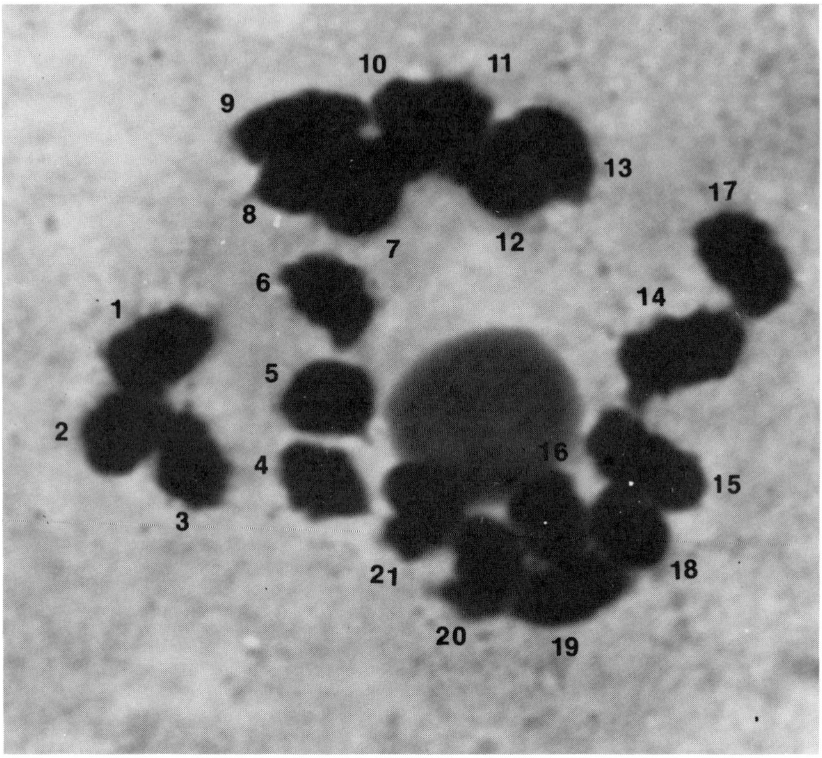

FIGURE 14. Metaphase I spread of haploid tall fescue displaying 14 unpaired (univalent) chromosomes and a group of 7 chromosomes which do not permit clear interpretation.

to support his hypothesis. Additionally, Jauhar points to the report from Malik and Tripathi[18] concerning the degree of chromosome pairing in the haploid of tall fescue studied by them as supportive evidence for the existence of a hemizygous ineffective pairing control mechanism.

However, the very low frequency of pairing found in the haploids examined by Eizenga and Kasperbauer[19] does not conform to a hemizygous ineffective pairing system functioning in tall fescue. Rather, the low frequency of pairing observed in this study is suggestive of either a hemizygous effective pairing control mechanism or extreme differentiation between the component chromosome sets that make up tall fescue. Again, we face the challenge of having only two studies which address a basic question, and the two studies present us with contrasting results. And, again, we are left with the obvious, albeit frustrating, problem of differences in the genetic background of the source material utilized in each of the studies. The problem may best be resolved by the use of haploids as a means of producing monosomic derivatives, as reported by Eizenga and Kasperbauer.[19] They reported that among

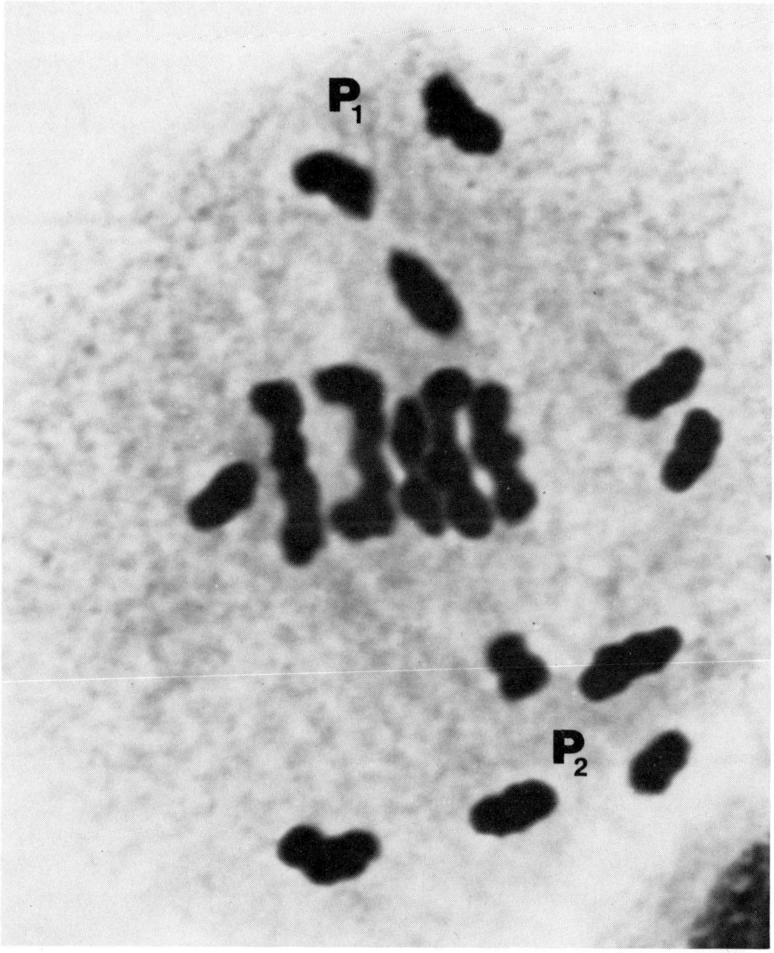

FIGURE 15. Late metaphase I/early anaphase I of haploid tall fescue showing 5 rod bivalents at the metaphase plate and 11 univalents. The univalents are unevenly distributed relative to the poles (P_1 and P_2) with three univalents still at the metaphase plate and three drifting toward one pole and five toward the other.

the progeny recovered from the open pollination of haploids, it is possible to recover aneuploids, some of which are monosomic. Careful study of a sufficient number of monosomics recovered by this means could confirm Jauhar's observations. It would require verification of each monosomic, by chromosome identification, in order to properly test the hypothesis. It is immediately obvious that a full set of monosomics would be required for the study.

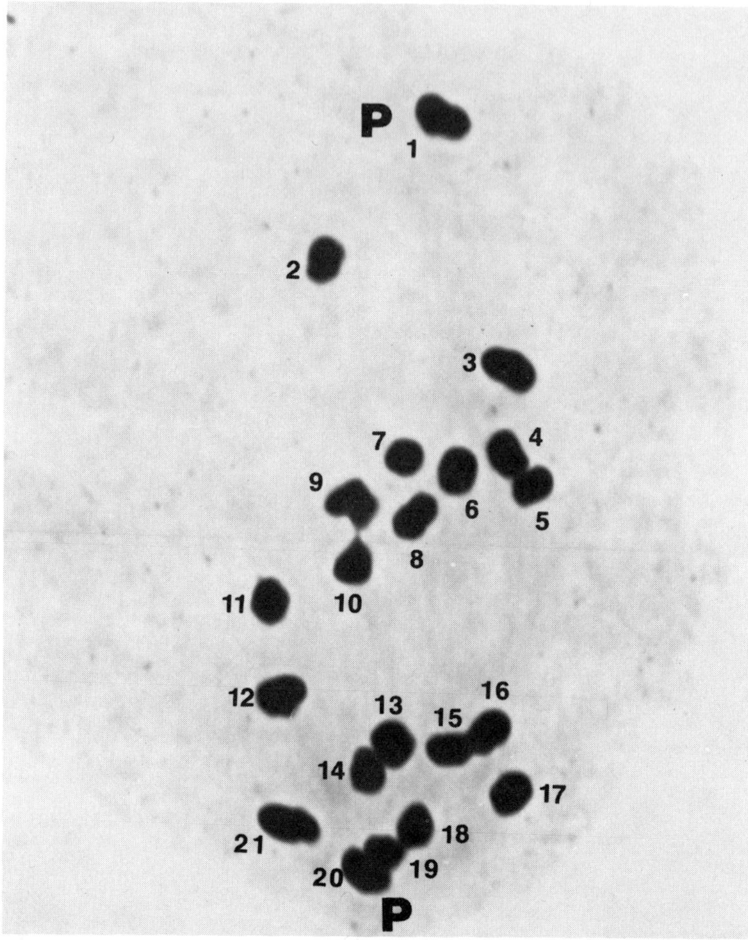

FIGURE 16. Typical mid-anaphase I of haploid tall fescue showing a very unequal distribution of the chromosomes following metaphase I. All 21 chromosomes of the haploid are clearly evident. The spindle poles are marked by the letter P.

As indicated above, the genetic nature of pairing control has significance to any breeding program concerned with the introgression of exotic genetic material into tall fescue. The more strictly that pairing is regulated (i.e., the greater the required homology for effective pairing), the less likely that desirable recombination events will occur. However, if in fact a hemizygous ineffective pairing system exists in tall fescue, the use of the appropriate monosomic, as female parent, in wide crosses would increase the probability of recombination between homoeologous chromosomes. Progeny from the monosomic amphiploid derivatives of such crosses may be the best source

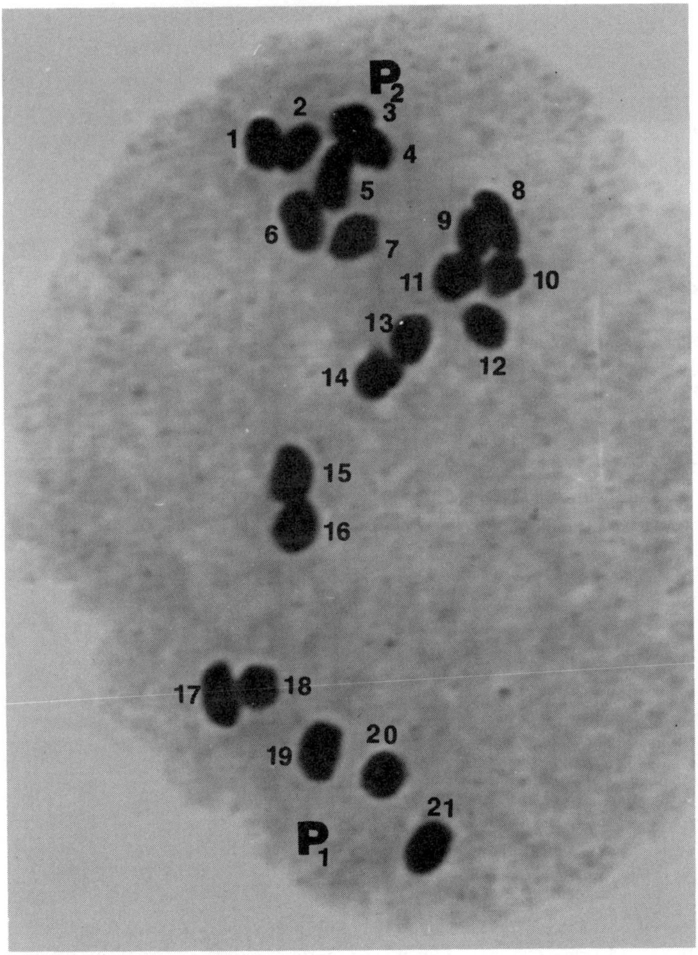

FIGURE 17. Mid-anaphase I of haploid tall fescue showing a distribution of chromosomes of 5 at one pole (P_1) and 14 at the other pole (P_2), with two chromosomes still at the spindle plate.

for new genetic combinations. The lack of a hemizygous ineffective pairing control mechanism would indicate that an equal probability for recovery of a particular desired recombinant is the same from all F_1-derived amphiploids. The existence of pairing control genes in hybridizing species would also be of significance.

In addition to the utilization of haploids in *in vitro* selection schemes, and to study the question of the nature of the control of chromosome pairing in tall fescue, haploids can also be useful to examine the apparent high degree

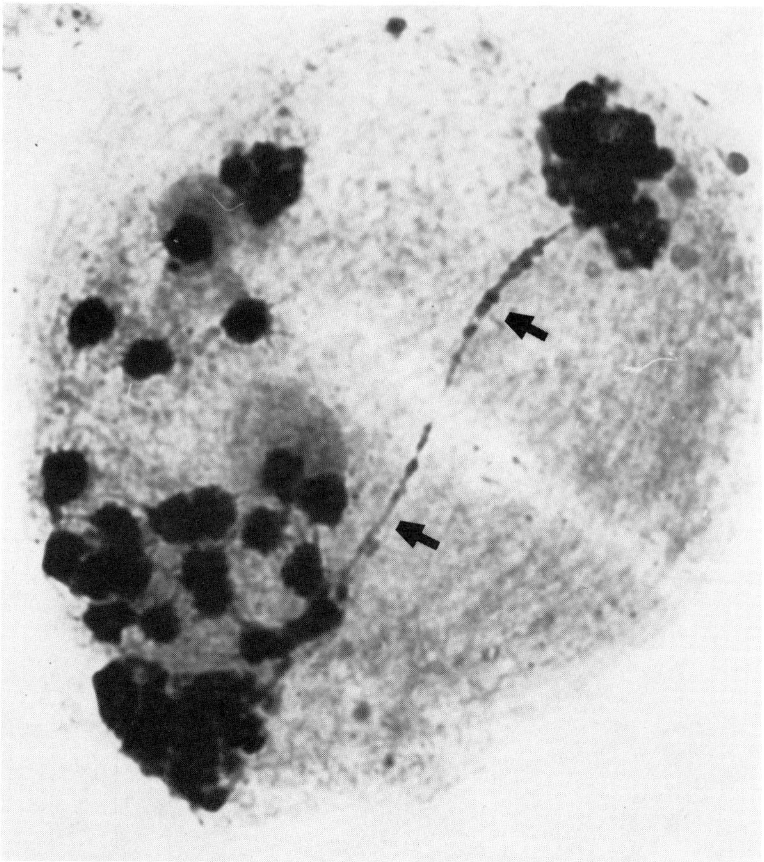

FIGURE 18. The occurrence of a chromatin bridge (arrows) between telophase nuclei in a telophase I cell. (The telophase cell is superimposed on a second cell which is in metaphase I.)

of genetic diversity in tall fescue. The results of Bulinska-Radomska and Lester[20,21] in their examination of seed proteins by the use of polyacrylamide gel electrophoresis adequately demonstrate the diversity found between populations of tall fescue. No indication of the degree of diversity within populations is indicated in these studies. Buckner[26] states that within the population known as Kentucky 31 there is a high relative degree of diversity. Seed from individual dihaploid derivatives of haploid plants could be examined by the same procedure employed by Bulinska-Radomska and Lester. The first question that this approach would answer is how much diversity is carried by individual plants relative to how much is carried within a population. It may also be possible to examine the degree of divergence between the component chromosome sets of the tall fescue genome.

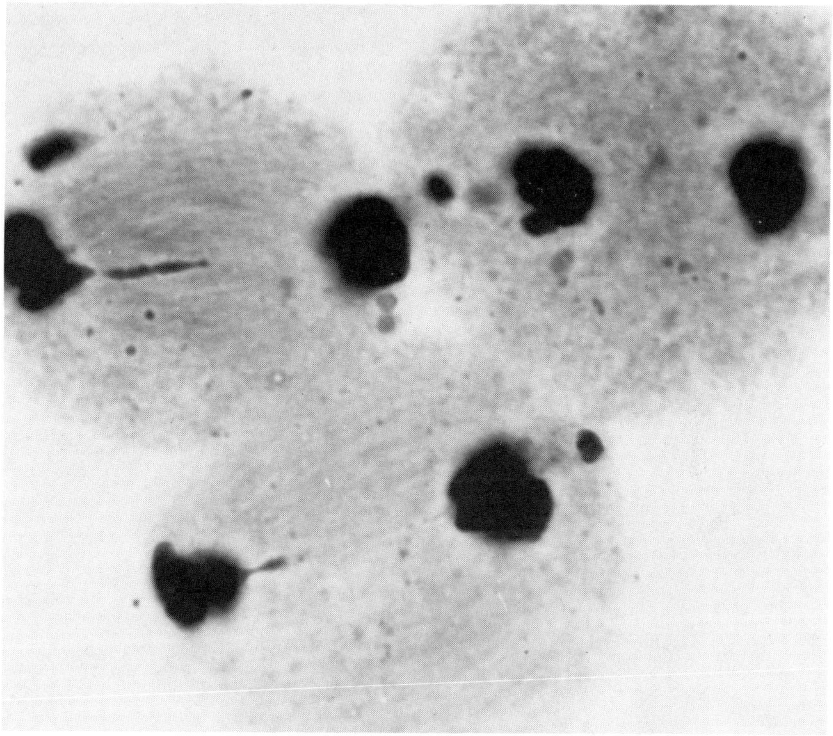

FIGURE 19. Evidence of chromatin bridges in several telophase I cells. Micronuclei can also be seen in these cells.

Finally, the ability to produce haploids consistently in sufficient numbers can be useful in mapping and linkage studies. This can be of real benefit in the case of polyploids.

C. Breeding Programs

Haploids, per se, have little utility in a breeding program. Breeding programs require the utilization of material which is reproductively compotent. Haploids are, at a minimum, male-sterile and, at best, may be weakly female-fertile in the case of certain polyploids, including tall fescue.[19] Therefore, other than using haploids for selection, as discussed above, haploids themselves are not useful in breeding programs. However, the dihaploids and other derivatives which can arise from haploids can be quite useful in various breeding strategies.

1. Hybridization

The first utility of haploid derivatives is the obvious applications which can be made of inbreds. The euploid derivatives of haploids which can be

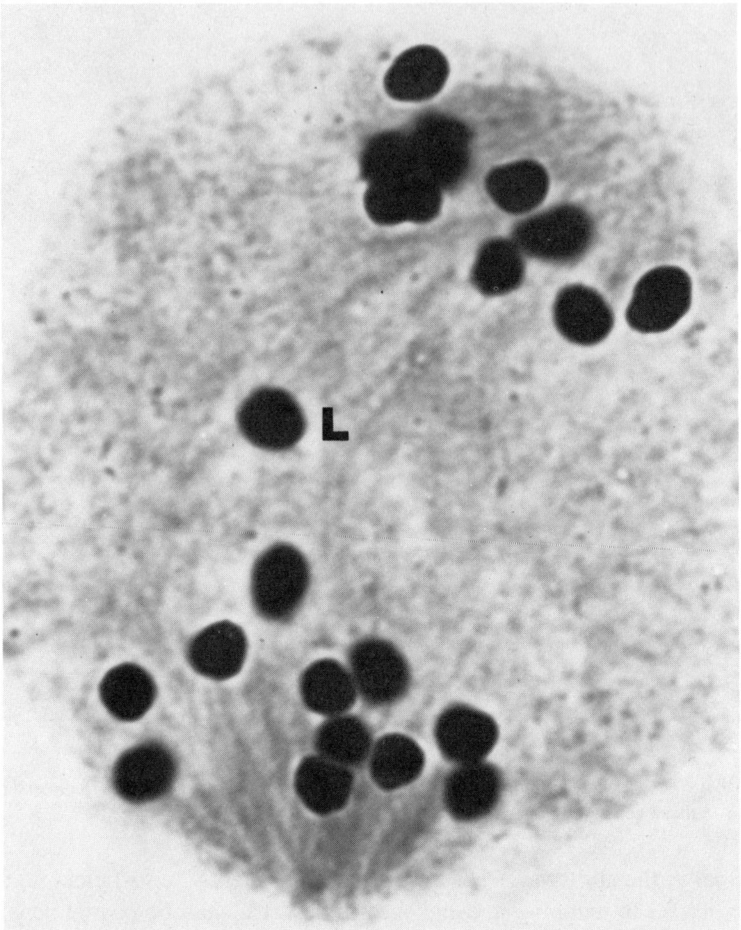

FIGURE 20. Anaphase I of haploid tall fescue showing a distribution of 9 chromosomes to one pole and 11 to the other pole, with 1 chromosome (L) apparently lagging at the spindle plate.

generated by chromosome doubling (either spontaneous or induced) would be inbred. In the case of quantitatively inherited characteristics, a breeding scheme which includes selection at the haploid level followed by chromosome doubling can provide a convenient mechanism for rapid consolidation of desirable genes for the characteristic. It is obvious that the characteristic being selected must be equally expressed in the haploid and the dihaploid in order for selection to be valid and useful. Multigenic disease resistance is an example of a characteristic which should fit such a program well, while seed yield,

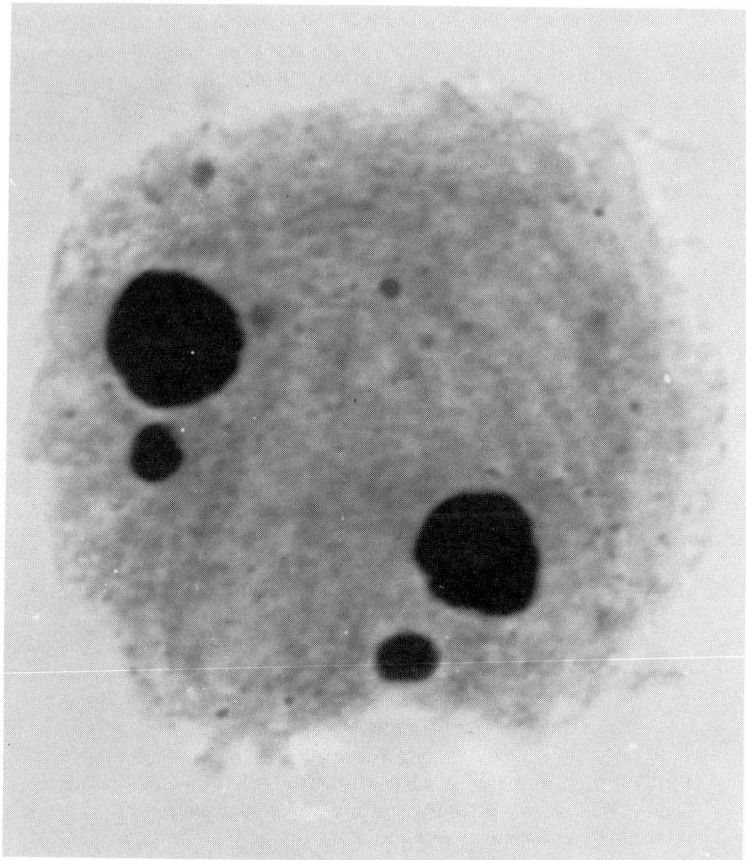

FIGURE 21. Telophase I of haploid tall fescue showing a relatively large micro-nucleus near each telophase nucleus.

which is a quantitatively inherited trait, is an example of a characteristic which is not amenable to such a system.

This general scheme could be used to move multigenic disease resistance from one breeding line to another, as an example. By generating the F_1 between a resistant line and a susceptible line, one can select for resistance among the gamete population of the F_1, as represented by the haploids derived thru anther-panicle culture. The resistant haploids can have their chromosome number doubled, and the subsequent fertile dihaploids can then be backcrossed to the recurrent (suseptible) line. The backcross plants would again be anther-panicle cultured and the subsequent haploids selected for resistance. This process should be more time efficient than a conventional backcross program.

The discovery by Eizenga and Kasperbauer[19] that tall fescue haploids do

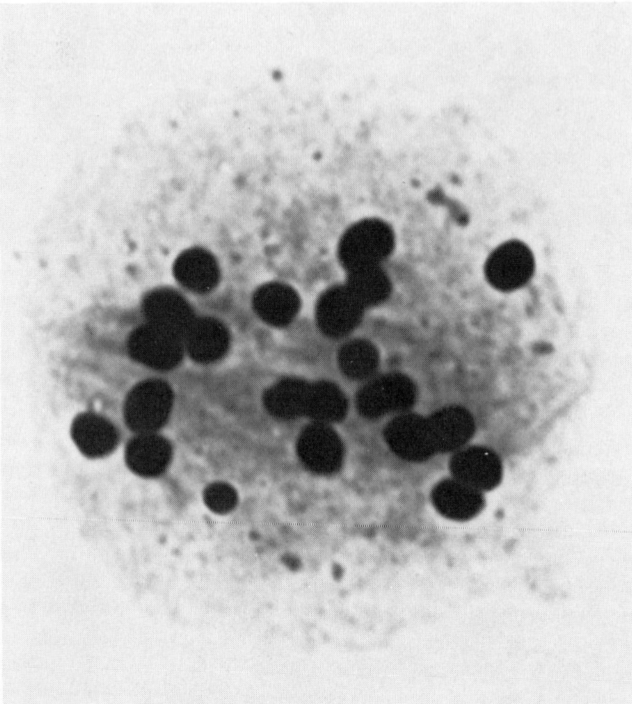

FIGURE 22. Anaphase I of haploid tall fescue showing 22 chromatin bodies. The presence of 22 chromatin bodies is suggestive of precocious division of a univalent at this stage.

exhibit a low degree of female fertility and that the progeny from the pollination of tall fescue haploids include various aneuploids, as well as euploids, offers a number of possibilities for utility in breeding programs. The recognition that monosomics are a significant part of the recovered progeny population from the fertilization of haploids is a very important discovery. Monosomics can be used to develop substitution lines. Substitution lines have been very useful in the breeding of several crops. However, the development of this opportunity will require the ability to differentiate the individual chromosomes of tall fescue better than is currently the case. The process of producing substitution lines is simply that of crossing whatever material we want to extract a particular chromosome from to the appropriate tall fescue monosomic, as female parent, and selecting F_1 progeny having a chromosome number equal to one less than the combined haploid complements of the two parental lines. After chromosome doubling, the amphiploid can be backcrossed, as pollen parent, to the monosomic, and individuals having a chromosome number equal to 20 plus one half the chromosome number of the

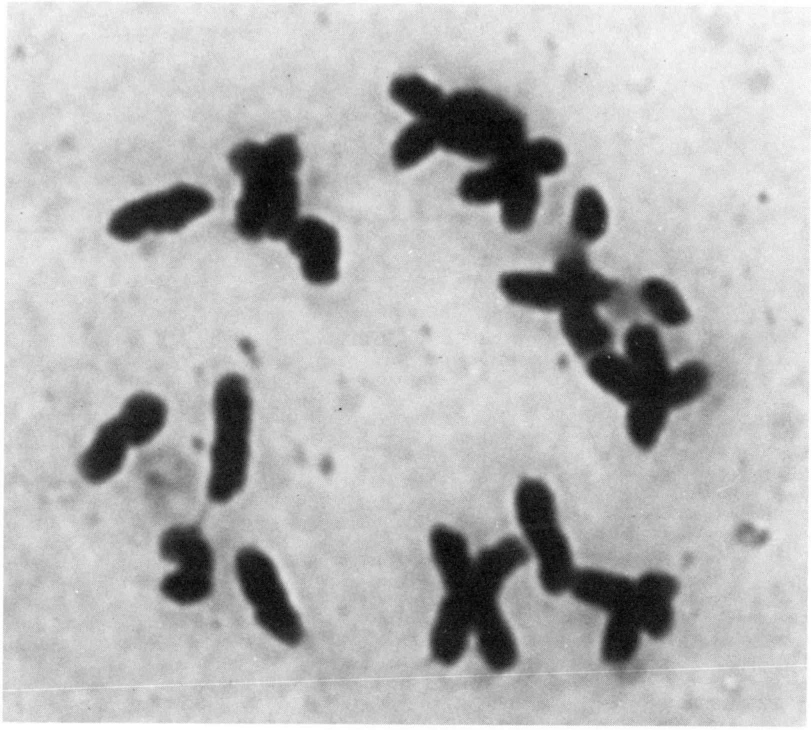

FIGURE 23. A single prophase II nucleus displaying seven chromatid pairs and seven unpaired chromatids. The unpaired chromatids are likely the result of precocious division of univalents at anaphase I.

amphiploid genome count are selected. These backcross progeny are then selfed, and chromosome numbers are monitored with selection for decreasing chromosome counts until we have recovered plants having a chromosome number of 42. These plants will have a chromosome complement comprised of 40 tall fescue chromosomes and 2 chromosomes from the second parent. The process is diagrammatically presented in Figure 24. By this process individual chromosomes can be replaced.

2. *Introgression*

The introgression into any species of characteristics from related germplasms is largely a function of the probability of recombination between homoeologous chromosome segments. Anything that the breeder can do to enhance the degree of pairing between homoeologous chromosomes would increase the effectiveness of attempts at introgression. As discussed above, a clear understanding of the genetics of pairing control would contribute

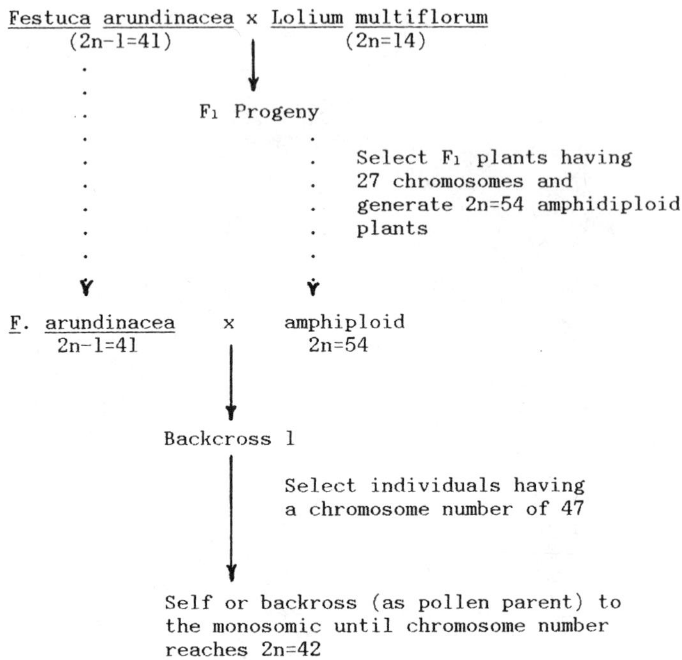

FIGURE 24. Diagrammatic scheme for utilizing tall fescue monosomics, derived from fertilization of haploids, to recover chromosome substitution lines. The use of *Lolium multiflorum* in this diagram is simply representative. Any cross-compatible species could be utilized in place of *L. multiflorum*.

strongly to this. Haploids will be a critical component in the unraveling of the confusion which currently surrounds this issue. Additionally, the prudent use of aneuploids in an introgression program can be very useful.

REFERENCES

1. **Rieger, R., Michaelis, A., and Green, M. M.,** *A Glossary of Genetics and Cytogenetics,* Springer-Verlag, New York, 1968,1.
2. **deWet, J. M. J.,** Origins of polyploidy, in *Polyploidy, Biological Relevance,* Lewis, W. H., Ed., Plenum Press, New York, 1979, chap. 1.
3. **deFossard, R. A.,** Terminology in "haploid" work, in *Haploids in Higher Plants, Advances and Potential, Proc. First Int. Symp.*, Kasha, K. J., Ed., University of Guelph, Guelph, Ontario, 1974, 403.
4. **Kimber, G. and Riley, R.,** Haploid angiosperms, *Bot. Rev.*, 28, 480, 1963.
5. **Sadasivaiah, R. S.,** Haploids in genetic and cytological research, in *Haploids in Higher Plants, Advances and Potential, Proc. First Int. Symp.*, Kasha, K. J., Ed., University of Guelph, Guelph, Ontario, 1974, 355.

6. **Crane, C. F., Beversdorf, W. D., and Bingham, E. T.,** Chromosome pairing and associations at meiosis in haploid soybean *(Glycine max), Can. J. Genet. Cytol.,* 24, 293, 1982.
7. **Malik, C. P. and Thomas, P. T.,** Cytological relationships and genome structure of some Festuca species, *Caryologia,* 20, 1, 1967.
8. **Jauhar, P. P.,** Genetic control of diploid-like meiosis in hexaploid tall fescue, *Nature (London),* 254, 595, 1975.
9. **Jauhar, P. P.,** Genetic regulation of diploid-like chromosome pairing in the hexaploid species, *Festuca arundinacea* Shreb. and *F. rubra* L. (Gramineae), *Chromosoma,* 52, 363, 1975.
10. **Malik, C. P. and Thomas, P. T.,** Chromosomal polymorphism in *Festuca arundinacea, Chromosoma,* 18, 1, 1966.
11. **Morgan, W. G.,** A technique for the production of polyploid grasses, *Euphytica,* 25, 443, 1976.
12. **Kasperbauer, M. J., Buckner, R. C., and Springer, W. D.,** Haploid plants by anther-panicle culture of tall fescue, *Crop Sci.,* 20, 103, 1980.
13. **Kasperbauer, M. J. and Eizenga, G. C.,** Tall fescue doubled haploids via tissue culture and plant regeneration, *Crop Sci.,* 25, 1091, 1985.
14. **de Lautour, G.,** A leaf squash technique for chromosome studies in grasses, *N. Z. J. Sci.,* 3, 293, 1960.
15. **Bennett, E.,** A rapid modification of de Lautour's technique for grass leaf chromosomes, *Euphytica,* 13, 44, 1964.
16. **Rogers, H.,** Annual Report Plant Breeding Institute, 1962—63, Cambridge, 1964, 89.
17. **Malik, C. P.,** Hybridization of *Festuca* species, *Can. J. Bot.,* 45, 1025, 1967.
18. **Malik, C. P. and Tripathi, R. C.,** Mode of chromosome pairing in the polyhaploid tall fescue (*Festuca arundinacea* Shreb 2n = 42), *Z. Biol.,* 116, 332, 1970.
19. **Eizenga, G. C. and Kasperbauer, M. J.,** Chromosome pairing in tall fescue haploids derived by anther-panicle culture, *J. Hered.,* 76, 99, 1985.
20. **Bulinska-Radomska, Z. and Lester, R. M.,** Relationships between three species of *Festuca* sect. Bovinae (Poaceae), *Pl. Syst. Evol.,* 149, 135, 1985.
21. **Bulinska-Radomska, Z., and Lester, R. M.,** Phylogeny of chromosome races of *Festuca arundinacea* and *Festuca mairei* (Poaceae) as indicated by seed protein electrophoresis, *Plant Syst. Evol.,* 152, 153, 1986.
22. **Malik, C. P.,** Pachytene studies in the intraspecific hybrid of *Festuca arundinacea* Shreb. (2n = 42), *Genet. Pol.,* 11, 379, 1970.
23. **Malik, C. P. and Thomas, P. T.,** Karyotypic studies in some *Lolium* and *Festuca* species, *Caryologia,* 19, 167, 1966.
24. **Alonso, L. C. and Kimber, G.,** The analysis of meiosis in hybrids. II. Triploid hybrids, *Can. J. Genet. Cytol.,* 23, 221, 1981.
25. **Kasha, K. J., Ed.,** *Haploids in Higher Plants, Advances and Potential,* Proc First Int. Symp., Kasha, K. J., Ed., University of Guelph, Guelph, Ontario, 1974.
26. **Buckner, R. C.,** Personal communication, 1980.

Chapter 8

PHYSIOLOGY AND BIOCHEMISTRY: POTENTIAL USE OF HAPLOIDS AND DOUBLED HAPLOIDS

C. J. Nelson and D. A. Sleper

TABLE OF CONTENTS

I. INTRODUCTION

The use of haploids and doubled haploids has great potential in understanding the physiological and biochemical processes that regulate growth of tall fescue (*Festuca arundinacea* Schreb.). Further, use of haploids and doubled haploids offers a potential means for more rapid development of cultivars that combine selected physiological traits. Conventional approaches are slow because tall fescue is an outcrossing polyploid, a species that has a majority of additive gene action, making it difficult to combine several desired physiological characters into a single cultivar.

Thus far, little information is available on the actual use of haploids and doubled haploids in this species for crop improvement. However, there are a number of physiological processes that may be altered genetically and incorporated into cultivars using these techniques. This chapter will concentrate on physiological characters that are under genetic control, some of our research toward better understanding of physiological processes associated with tall fescue productivity, and potential use of haploids and doubled haploids to accelerate incorporation of improved physiological characteristics into new or improved cultivars.

II. PHYSIOLOGY OF ADAPTATION

Tall fescue is adapted to a wide range of climates, especially temperature and, to a lesser extent, rainfall.[1] Some strains are more winterhardy than orchardgrass (*Dactylis glomerata* L.), but are less winterhardy than bromegrass (*Bromus inermis* Leyss.) and timothy (*Phleum pratense* L.). Tall fescue is grown generally where winter temperatures tend to be mild to exploit its genetic potential for low-temperature growth. For example, Leasure[2] reported that tall fescue grew when weekly temperatures were above 4.4°C, whereas orchardgrass grew very little unless weekly temperatures averaged over 10°C. During late fall and winter, leaf blades of tall fescue remain green and photosynthetically active after being exposed to air temperatures as low as −10°C, allowing the species to produce and retain high-quality herbage for winter grazing in mid latitudes of the U.S.

Similar to other cool-season grasses that utilize the C_3 photosynthetic pathways, tall fescue produces about 65% of its annual herbage yield during spring. In contrast, warm-season grasses which utilize the C_4 photosynthetic pathway produce a majority of their herbage yield during the summer months if moisture is available.[3] Tall fescue does not grow rapidly in the warm (average daily temperature of 25 to 30°C), dry summers in the east-central part of the U.S. It was noted early, however, that tall fescue grows well during autumn, when herbage yield is predominately leaf blades. This led many farmers to develop management systems whereby tall fescue is allowed to grow ungrazed from mid summer into the autumn (called stockpiling), then grazed during much of the winter when little net increase in dry matter occurs.

Most tall fescue breeders are interested in improving yield during the vegetative stage to improve summer and, especially, fall production. Thus, much of our research on tall fescue has focused on photosynthesis-yield relationships, the tillering process, leaf growth rates, and herbage quality of vegetative canopies. Long-term objectives are to genetically improve forage productivity during summer and fall.

A major attribute in the adaptation of tall fescue is its larger root structure compared with other cool-season forage grasses. Secondly, tall fescue root systems distribute deeper in the soil than those of grasses such as orchardgrass. Because of these attributes, tall fescue adapts better than other C_3 species to water stress environments and to fluctuating water tables.[4]

Seedling establishment can be a problem with tall fescue and has been addressed physiologically.[5] Several reports show there is genetic variation for seed germination and seedling vigor in tall fescue. Further, there are known allelopathic effects on other seedlings due to residue from tall fescue,[6] and genotypic differences in allelopathic potential have been demonstrated.[7] The exact chemical composition of the inhibitory compound has not been clearly described.[8] In this chapter, however, we would like to emphasize the potential for improving physiological aspects of herbage yield of tall fescue in established swards.

Historically, tall fescue has not compared favorably with other cool-season grasses in terms of animal performance.[9] However, with the better understanding of the endophyte fungus problem (see Chapter 1 of this book), a great potential exists for improved animal performance. At the same time, there is a growing realization that the endophytic fungus may have coevolved in a manner to give infected tall fescue plants improved persistence and competitiveness.[10] Factors such as disease resistance, insect resistance, and nematode resistance may be augmented by presence of the endophyte; therefore, special consideration at the physiological and biochemical levels should be given to the benefits and liabilities of the endophyte. For example, West et al.[11] found fewer soil-borne nematodes on roots of endophyte-infected tall fescue, which conferred better drought tolerance compared with noninfected plants.

Tall fescue, by its seasonal nature of growth patterns and its utilization by ruminants, provides a unique research opportunity for crop physiologists. While herbage yield is relatively easy to define or measure, the manner by which yield is developed is not completely understood. Improved knowledge of the mechanisms of canopy development and their genetic control should allow breeders to be more efficient in yield improvement.[12] At the same time, using physiological information to improve only yield would appear simplistic, when physiological factors associated with forage quality and plant persistence also need to be considered.

III. PHYSIOLOGY OF HERBAGE PRODUCTION

Three major concepts have evolved that describe forage yield in terms of solar energy capture and use. These include (1) rapid development of the leaf canopy to maximize light interception, (2) improvement of the leaf carbon budget by increased photosynthesis and/or decreased dark respiration, and (3) use of management systems based on the growth habit of the plant to maintain an adequate stored energy reserve. Each of these will be considered separately and then integrated to suggest ideotypes.

A. Light Interception

1. Canopy Structure

The concept of "critical" leaf area index (LAI) was developed by Brougham[13] to describe the LAI necessary to achieve 95% light interception at noon. This concept was researched in terms of forage management to maximize radiation interception throughout a growing season.[14,15] We have determined the critical LAI for a vegetative canopy during fall to be about 6.5 for "Kentucky 31" tall fescue in Missouri, a value that compares favorably with most cool-season or C_3 forage grasses. Critical LAI depends on size and orientation of leaves and would need to be described for each species. For example, we have selected populations of tall fescue having rapid and slow leaf elongation rates,[16] which in turn differ in size and orientation of leaves. Extinction coefficients for light penetration differ due to canpoy shape. Rapid leaf elongation rate populations usually have their leaf blades oriented more vertically, allowing better light penetration.

Forage grasses tend to intercept a rather small proportion of the solar energy available during a growing season because they often do not reach the critical LAI. Hart et al.[17] reported that when tall fescue was cut weekly, even while retaining a 20-cm stubble height, only 33% of the total radiation during the growing season was intercepted. It is obvious that management systems and new cultivars need to be developed to increase light interception. We evaluated light interception during the season, March 15 to October 15, for several genotypes of tall fescue that were fertilized with 0, 90, 180, or 270 kg/ha of nitrogen annually. Plots were cut and nitrogen was applied four times annually. Nitrogen applications were proportional to the percentage of annual forage produced during each growth period. Averaged over 2 years, the integrated light interception for the season was about 32, 51, 61, and 64%, respectively, of the total.[18] Plants were cut about 6 cm, and by the end of most regrowth periods, especially in vegative stages, the plants had not reached 95% light interception.

Rhodes[19] evaluated canopy structure of forage grasses and its relationship to productivity. He concluded that the desired canopy shape would be a function of the management system for which it was intended. For example, with infrequent cutting, which usually gives the highest forage yield, canopies should have long and rigid leaves with erect tillers. Rhodes[20] has shown that genetic variation exists for leaf length in perennial ryegrass (*Lolium perenne* L.) populations. Later, he selected populations that were either long or short leafed. He found that yield of the long-leafed population was superior with infrequent cutting, but yield of the short-leafed population was superior with frequent cutting.[21] Rhodes and Mee[22] compared several morphological features and suggested that leaf length had the largest effect on herbage productivity, but it was essential that the long leaves be held upright by increases in both tiller angle and leaf rigidity.

2. *Leaf Elongation Rate*

While leaf length is an important factor, we felt that leaf elongation rate of tall fescue may be a better selection index.[23] Leaf length is correlated positively with leaf elongation rate, and both length and elongation rate are correlated positively with leaf width.[16] In addition, there is ample evidence that leaf length is partially influenced by duration of the elongation period, i.e., the time interval between appearances of successive leaves,[24-26] suggesting that long leaves may be influenced by factors other than elongation rate. Nevertheless, to achieve rapid increases in leaf area on a canopy basis, we would want leaf elongation to be rapid.

Brougham[14] and Booysen and Nelson[27] suggested that pasture productivity would increase if plant canopies were managed to be near their critical LAI to maximize radiation interception. This management strategy would maximize herbage yield. More recently, however, research with perennial ryegrass suggested that when pastures are managed at a high LAI, although below the critical LAI, animal product per unit of land is not maximized.[28] Instead, the data suggest that closer grazing, to maintain a lower LAI, is beneficial because close grazing minimizes the number of leaves that decay and die in the lower canopy. With lax grazing, death and decay of lower leaves can be as much as 40% of the herbage produced.[29] However, when pastures are grazed to maintain a lower LAI, the animals consume a higher percentage of the forage and the quality of the forage is better. Therefore, the increased utilization tends to offset the slightly slower growth rate of the sward due to the canopy having less than the critical LAI.

When canopies are managed at LAIs below the critical LAI, the rate of leaf growth becomes very important. Leaves are the primary photosynthetic organs, and an increase in LAI will increase light interception and sward growth rate. Under these conditions it is important that rapid leaf growth occurs without a lengthening of the plastochron (time interval between successive leaf initiations), because axillary sites must be produced to allow development of lateral tillers to maintain good ground cover.[30] Number of axillary sites produced is directly related to the number of leaves produced.

Since our earlier work suggested that leaf net photosynthesis was generally not correlated positively with herbage yield,[31] we sought to determine the reasons for this effect.[32] Our conclusion was that in addition to having high net photosynthesis per unit leaf area, the plant should have an active sink to utilize the photosynthate. Placing selection pressure on leaf elongation rate or leaf area expansion rate (leaf elongation rate multiplied by leaf width) may influence the amount of photosynthate partitioned into leaf growth. When sink demand of the economic product (leaf blade) is increased, the high net photosynthetic rate of leaves may support the sink and improve yield.

3. Regulation of Leaf Growth

Leaf area expansion rate is critical because leaves are the harvestable product, and they contribute to an ever-increasing LAI, which in turn increases the size of the light intercepting surface of the canopy until the critical LAI is reached. We have found significant genetic variation for leaf area expansion rate of tall fescue.[33] A genotype with a faster rate of elongation produced more cells per day in the leaf meristem, and the cells expanded to a greater length when compared to a genotype exhibiting slow leaf elongation rate.[34] The greater cell length was associated with a longer duration of cell enlargement in the growth zone, perhaps due to longer sheaths in rapid leaf growth genotypes which protect the elongation zone from red light (see Chapter 4 of this book for discussion of the role of far-red/red light ratio via phytochrome on autoregulation of tillering). Based on our research which showed that leaf width was associated positively with elongation rate, we began a recurrent selection program for leaf area expansion rate (see Chapter 9 of this book).

Soil moisture and water stress have a large effect on the rate of leaf elongation.[35] We studied diurnal effects on leaf elongation and found that with constant meristem temperatures the leaves elongate faster at night.[36,37] Further, differences in leaf elongation rate between genotypes may be greater during night than during the day. Compared with the genotype selected for slow leaf elongation, the rate for a high leaf elongation genotypes was 40% higher during the photoperiod, but 57% higher during the dark period. Leaf water potential for both genotypes was lower during the photoperiod than during the dark period. The genotype with high leaf elongation rate had a higher leaf-water potential during the photoperiod, and especially during the dark period, suggesting that it had a more favorable water balance than the genotype with low leaf elongation.[36] The more favorable water balance should facilitate cell growth.

We have shown that the leaf elongation zone contains large amounts of water-soluble carbohydrates (WSC), mainly fructan.[38] Further, fructan in the leaf growth zone of tall fescue has a low degree of polymerization,[39] allowing it to contribute osmotically to the water potential of growing cells. Schnyder and Nelson[40] reported about 0.4 MPa of the osmotic potential in the cell elongation zone was due to water-soluble carbohydrates. About half was due to fructan, and about half to sucrose and monosaccharides.

At this time it is not clear what regulates leaf area expansion rate, but we believe the sink is not carbohydrate limited.[40-42] We have shown that plants maintaining a more positive water balance have more rapid leaf growth, but this general observation needs verification by studies using elongating cells. When we treated plants with gibberellic acid (GA_3), we obtained a marked increase in leaf elongation rate, but the leaves also were narrower and thinner.[43] We did not evaluate cell number or cell size in this study, but the data suggest that processes associated with wall loosening (to allow cell expansion) should be considered. Other data show clearly that cell production rate is more responsive to nitrogen application than is cell elongation rate.[44,45] This suggests that some attention needs to be given to processes that regulate the cell division process.

4. Tillering and Weight per Tiller

Tillering activity plays a critical role in the relationship between growth and yield. In spaced-plant conditions, we showed that rate of tillering is about 1.5 times more critical than weight per tiller in determining yield per plant.[12] In sward conditions, however, Hart et al.[17] found that tiller density of tall fescue tends to stabilize for a given management system and remains nearly constant during the vegetative growth period. When tiller density is stable, yield is determined largely by weight per tiller. Weight per tiller is related closely to leaf area expansion rate.[12]

Zarrough et al.[46,47] actively pursued the relationships between yield and tiller density of tall fescue genotypes. Genotypes selected for low yield per tiller tended to have higher tillering rates than those selected for high yield per tiller when vegetative plants were transplanted to the field. As tiller density increased to close the canopy, death of smaller tillers began to offset the continued production of new tillers, causing net number of live tillers per unit of land area to stabilize. Until tiller density stabilized, the rate of tillering and subsequent density of tillers were the most critical yield determinants. Yield per tiller became the most critical yield determinant after a stable tiller density was reached.

Responses of genotypes selected for low and high yield per tiller were compared under frequent (every 2 weeks) and infrequent (every 6 weeks) cutting management.[48] Invariably, genotypes selected for low yield per tiller had high tiller density, had large amounts of residual leaf area after cutting, and yielded better under frequent cutting. In contrast, when cutting was infrequent, the genotype with high yield per tiller was able to develop long leaves and have yield superiority over the genotype with low yield per tiller. This is consistent with data for perennial ryegrass.[19]

B. Leaf Carbon Budget

In the broad sense, net photosynthetic rate (carbon dioxide exchange rate, CER) is a major yield determinant. Therefore, the efficiency by which a crop canopy converts solar energy to usable organic forms through photosynthesis is an important consideration in forage breeding and management. Genetic variation exists for CER in tall fescue.[49] However, several researchers have shown a lack of a relationship between leaf CER and forage yield.[50] For example, Rhodes[19] and Sheehy and Cooper[51] reported a poor and often negative relationship between leaf CER and yield of ryegrass and other grass species during summer. Also, Hoveland et al.[52] reported no relationship between winter growth and cool-temperature CER in *Phalaris tuberosa* L. and tall fescue. We found a low and inconsistent correlation between forage yield and CER on a leaf area basis among several genotypes of tall fescue measured in the field.[50,53]

While the poor relationship between CER and herbage yield is perplexing, it is not entirely irrational, as several factors including mesophyll cell size, environmental stress, dry matter distribution, and genotype by environment interactions can all affect the relationship (see discussion in Nelson[50] and Nelson et al.[53]). A probable reason for the poor association is that the LAI is often not large enough to intercept all of the light. Thus, the amount of leaf area rather than the CER per unit leaf area tends to be the major yield-limiting factor. Successive leaves appearing on grass plants in vegetative canopies usually have lower CER rates than their predecessors because they develop in a more shaded environment.[54] Therefore, canopies with long leaves and high productivity have leaves developing at the base of the plant in a low-light environment that predisposes them to lower CER rates. Further, characters associated with rapid leaf elongation, such as large cells, are often associated with lower photosynthetic rates.[55] However, in our research with leaf elongation[16] and tall fescue polyploids,[56] we have not seen a strong negative relationship between CER and cell size.

Apparently it is going to be difficult to have both high leaf growth rate and high photosynthetic rate on a leaf area basis. Even so, at the University of Missouri, we are developing separate populations using recurrent selection for leaf area expansion rate and CER. After four cycles of selection for rapid and slow leaf area expansion rates, we have more than a twofold divergence, yet CER is not markedly different between the divergent populations.[16] Conversely, after four cycles of selection from the same base population, plants selected for high CER tend to have shorter leaves than those selected for low CER.[57] We do not know the reason for the difference in the response. Our present objective is to cross plants selected for high leaf area expansion rate with those selected for high CER. Our hypothesis is that the combination of an active sink (leaf meristem) with high CER will give further potential for selection of superior yielding types.

C. Carbohydrate Storage and Use

The concept of carbohydrate reserves and their storage and recycling needs to be considered in terms of the overall management of the grass crop. During periods when growth is not as rapid as photosynthesis, some storage will occur. Conversely, when photosynthesis is slowed due to defoliation or unfavorable climatic conditions, stored energy is used for supporting the new growth. While the exact role of stored energy has been questioned,[58] it is known to be involved in regrowth of tall fescue.[27,59,60]

We compared rates of vegetative regrowth of tall fescue for 20 d following pretreatment to give low and high levels of WSC in the stem base tissue.[27] Within each WSC treatment, we either left some leaf area on the cut stubble or removed all leaf blades. Regardless of concentration of WSC, the rate of dry matter accumulation of tops plus roots was rapid for plants with leaf area remaining, whereas regrowth rates with high WSC and low leaf area were considerably slower. This verifies the value of residual leaf area for rapid regrowth of cool-season grasses.[14,61,62]

Total energy storage in the plant must be interfaced with leaf growth rate and CER, as discussed in this chapter. The ultimate objective is to have canopies that quickly develop leaf area and have a high photosynthetic output. Our data with tall fescue indicate that residual leaf area is helpful in regrowth. Therefore, we have postulated that some leaf area should be left after cutting or grazing to intercept radiation and provide energy for regrowth.[14] This would be most critical under frequent defoliation, would decrease dependence of regrowth on reserve carbohydrates, and would allow more energy fixed by photosynthesis to remain in the top growth, where it could be utilized directly by grazing animals. However, when a tall stubble is left after cutting or grazing, some decay of leaf material occurs over time.[29] Thus, maximum expected benefit of residual leaf area during the entire season may not be achieved[17,28] unless the remaining material is "hard grazed" in autumn.[15]

Our research shows that populations selected for a low rate of leaf area expansion have lower concentrations of WSC in the stem bases than do those selected for a high rate.[59] This suggests that part of the reason the low leaf area expansion rate populations do better with frequent cutting than do the high leaf area expansion populations is that they are less dependent on WSC and are more dependent on residual leaf area to support regrowth. In contrast, the more upright canopy of the genotype with rapid leaf elongation has a lower tiller density and lower residual leaf area than a genotype with slow leaf elongation. If cut frequently, the high leaf elongation rate genotypes probably cannot restore high concentrations of WSC between cuttings and have a slow regrowth rate due to low residual leaf area.[27,59]

We studied leaf senescence of tall fescue to determine if it is possible to retain photosynthetically active leaf area in the stubble after cutting, but so far, results are not encouraging.[63] Apparently, the plant catabolizes protein and RuBP carboxylase from older, shaded leaves to support young, growing tissue, even when large amounts of nitrogen are added.

IV. PHYSIOLOGY OF POLYPLOIDS

Tall fescue is an outcrossing polyploid that behaves like a diploid forming

21 bivalents (2n = 6x = 42 chromosomes) at meiosis (see Chapter 9 of this book). The diploid behavior may facilitate use of molecular and haploid approaches to plant improvement. Another asset is that tall fescue will cross with a wide range of *Festuca* spp., thus allowing progeny from interspecific crosses to be effectively utilized. Further, tall fescue crosses readily with perennial ryegrass (*Lolium perenne* L.), giving a broader diversity of physiological characters with which to work.

A. CO_2 Exchange

Festuca arundinacea is part of a large, naturally occurring polyploid series of *Festuca* (see Chapter 9 of this book). We have examined extensively the photosynthesis relationships among the polyploids.[64] Our overall objective was to better understand the phylogenetic relationships of the *Festuca* genera and to evaluate the genetic regulation of photosynthesis. Our early work suggested a ploidy effect on photosynthesis because a decaploid genotype had a rate that was 58 to 74% higher than the average of several hexaploids.[65] We then selected genotypes solely on the basis of ploidy at each of four levels (4x, 6x, 8x, and 10x), with the genomic constitution of the 4x being G_1,G_1,G_2,G_2 (Table 1). Photosynthesis per unit of leaf area increased with ploidy.[66]

When the diploid *F. pratensis* (PP) and the tetraploid *F. mairei* ($M_1M_1M_2M_2$) were added to the data set, the direct ploidy relationship was less clear (Table 1). Resolution of ploidy effects on photosynthetic rates will be greatly aided by using different ploidies derived from the same genotype, such as the 21-, 42-, and 84-chromosome lines derived by induced endomitotic divisions at the cellular level and regeneration of plants of higher ploidy, as discussed in Chapter 6 of this book.

The observed correlation between net photosynthesis and content of RuBP carboxylase suggests that at least part of the increase is due to the amount of carboxylating enzyme.[66] The higher enzyme content should lead to a decreased mesophyll resistance to CO_2 uptake if the enzyme is fully or equally activated at all ploidy levels. Response of CER to a range of O_2 concentrations shows that photorespiration is closely associated with CER, and *in vivo* properties of RuBP carboxylase are not altered by ploidy level.[67]

To further evaluate the ploidy response, we evaluated stomatal density and mesophyll cell volume of plants growing in a controlled environment (Table 2). As in many studies using polyploids, stomatal density decreased and length increased as ploidy increased from 4x to 8x. Stomatal density was highest and stomatal length was greatest with the decaploid (2n = 10x = 70) plants, and both factors contributed to the low stomatal resistance. Several studies have established that stomata are highest in density on the adaxial surface of tall fescue. Mesophyll cell volume also increased with ploidy between 4x and 8x. Cell volume for the 10x genotypes was intermediate between the 4x and 6x genotypes, and stomatal density was similar to the 4x

TABLE 1
Relationship of Net Photosynthesis Rates (CER) and Associated Character among *Festuca* Ploidy Levels

Chrom. number	Genomes	CER[a]	SLW[b]	Soluble protein (mg/g fr. wt.)	RuBPc (mg/g fr. wt.)	RuBPc/ protein (%)
14	PP	20.8	5.1	42.2	6.8	16.7
28	$G_1G_1G_2G_2$	19.1	5.9	40.8	8.7	16.8
28	$M_1M_1M_2M_2$	28.7	8.7	32.5	9.6	30.8
42	$PPG_1G_1G_2G_2$	22.8	5.9	41.4	9.2	22.2
56	$G_1G_1G_2G_2M_1M_1M_2M_2$	30.1	5.7	37.2	10.1	26.2
70	$QQG_1G_1G_2G_2M_1M_1M_2M_2$	32.2	5.6	37.4	10.5	28.4

Note: Plants were grown in the field; CO_2 exchange was measured on the last fully developed leaf of vegetative plants at saturating radiation. Data are the means of two experiments.

[a] Net photosynthesis, milligrams of CO_2 uptake per square decimeter of leaf area per hour.
[b] Specific leaf weight, milligrams per square centimeter of leaf area.

Adapted from Randall, D. D., Nelson, C. J., Sleper, D. A., Miles, C. D., Crane, C. F., Krueger, R. W., Wong, J. H. H., and Poskuta, J. W., in *Nitrogen Fixation and CO_2 Metabolism*, Ludden, P. W. and Burris, J. E., Eds., Elsevier, New York, 1985, 409. With permission.

TABLE 2
Effect of Ploidy Level on Stomatal Density on the Adaxial Surface, Leaf Diffusive Resistances, and Net Photosyntheses (CER) of Tall Fescue Leaf Blades

		Stomata		Resistance		CER[a]		
Chrom. number	Genomes	Density (no./mm^2)	Length (μm)	Adax.	Abax.	g^{-1}	dm^{-2}	Cell vol[b]
28	$G_1G_1G_2G_2$	43.6	41.8	6.7	17.7	32.7	15.9	1.0
42	$PPG_1G_1G_2G_2$	36.3	45.5	2.4	5.5	34.6	18.8	6.2
56	$G_1G_1G_2G_2M_1M_1M_2M_2$	30.6	45.8	2.4	8.8	46.6	22.3	7.2
70	$QQG_1G_1G_2G_2M_1M_1M_2M_2$	47.2	50.7	0.8	3.7	44.4	21.7	4.7

[a] Photosynthesis rate expressed as milligrams of CO_2 uptake per hour on a leaf weight or area basis.
[b] Cell volume assumes the third dimension was proportional to width, and the 28-chromosome genotype had an index of 1.0

Adapted from Byrne, M. C., Nelson, C. J., and Randall, D. D., *Plant Physiol.*, 69, 891, 1981. With permission.

plants. Smaller mesophyll cell size has been associated with high net photosynthesis in perennial ryegrass.[55]

B. Light Reactions

Colleagues at the University of Missouri evaluated genetic variation and the ploidy influence on electron transport and adenosine triphosphate (ATP) synthesis in tall fescue.[68] Isolated chloroplasts were used to avoid problems with leaf morphology as described above. No correlation was found between ploidy level and electron transport for photosystem I (PSI), but one 10x genotype and a 4x genotype ($M_1M_1M_2M_2$) with high net photosynthesis (Table 1) had the highest PSI rates. The conclusion was that certain genotypes could display high PSI electron transport rates, probably involving the P700 reaction center. There were no general relationships, however, between ploidy level and light reactions of isolated chloroplasts for the genotypes used.

The chloroplast genome would not necessarily be expected to be altered by the nuclear ploidy level, but its regulation may be affected. Clearly, there is a need to know how the various levels of genetic regulation occur. It appears that electron transport is not the limiting factor in these genotypes. Overall, in tall fescue the metabolism associated with CO_2 fixation appears to be more limiting.[64] The dark reactions involving RuBP carboxylase and the myriad of enzymes associated with CO_2 reduction would involve many nuclear and chloroplast genes, making it difficult to manipulate genetically.

Collectively, one can conclude that the genetic regulation of photosynthesis is complex and can be influenced by many factors simultaneously. What are the most likely explanations? Some insight can be gained by reviewing the data in Tables 1 and 2. It appears that adding genomes can increase photosynthetic potential, and that the 4x *F. mairei* genome has a particularly strong influence. An increase in content of the primary carboxylating enzyme (Table 1) could work with the decreased diffusive resistance (Table 2) in a complementary way to enhance CO_2 uptake.

Genetic regulation might have coevolved for stomatal functions and carboxylation potential to be matched in a positive way. In subsequent research, Allard et al.[69] found that shading caused simultaneous reduction of carboxylation activity and stomatal frequency such that internal CO_2 concentration remained unchanged. Our data on cellular maturation during leaf growth suggests that stomata differentiation occurs a few days earlier in leaf development than does synthesis of RuBP carboxylase.[44] Subsequent experiments need to be conducted to determine if the processes can be separated genetically.

A basic physiological process such as photosynthesis is regulated genetically and can be influenced by ploidy level. Our current thinking is that with this operational allopolyploid the response is not due to ploidy per se, but to the frequency of desirable alleles being present. It is likely that natural selection in the progenitors had already occurred before extensive polyploidization had developed. This cannot be tested completely until the exact progenitors are known. Even so, the naturally occurring polyploid series gives a useful tool for studying photosynthesis. However, different ploidy levels derived from the same genotype (as described in Chapter 6 of this book) are needed to more clearly define ploidy effects under different environments.

V. PHOTOSYNTHESIS OF HAPLOIDS

Haploids (n = 5x = 35) of decaploid tall fescue were obtained from polyembryonic seed. Nearly 70,000 seeds were germinated at 20°C, and a low frequency of haploid plants was observed as the coleoptiles were emerging. The five polyembryonic seedlings were transplanted to pots, verified as haploids using root tip squashes, and later evaluated for photosynthetic potential and other characteristics.[70] For comparison, F_1 hybrids were made between the tetraploids *F. mairei* and *F. arundinacea* var. glaucesens. These F_1 hybrids had a genomic structure of $G_1G_2M_1M_2$, which is similar to that of the polyhaploids, namely $Q_1G_1G_2M_1M_2$ (see Table 1).

The CER of the F_1 hybrids (23.5 mg dm^{-2} h^{-1}) was intermediate between the two tetraploid parents, again verifying that inheritance is due largely to additive gene action. The CERs of the haploids ranged from that near *F. arundinacea* var. glaucesens (13.5 mg dm^{-2} h^{-1}) to that of the F_1 hybrids. Again, additive gene action is suggested since none of the haploids (12 to 24 mg dm^{-2} h^{-1}) approached the CER of *F. mairei* (30 mg dm^{-2} h^{-1}), which is similar to the decaploid from which they came (Table 1). If increasing ploidy per se is responsible for increased CER, we would expect the resulting haploids to be uniformly lower in CER, about 50% of the decaploid, assuming dark respiration is also affected proportionately by ploidy.

With the genotypes used, we showed that ploidy level did not influence the *in vivo* properties of RuBP carboxylase, as rates of photorespiration and CER were proportional for all ploidy levels.[64,67] Further, the haploids had CER responses to oxygen concentrations that were similar to other *Festucas,* suggesting that *in vivo* properties of RuBP carboxylase were not altered in the haploid state.[70] Additonal studies are planned using haploids and their doubled haploids, as used for other comparisons of ploidy effects within the same genotype.[71]

VI. GENETIC VARIATION IN PHYSIOLOGICAL CHARACTERS

Several important physiological traits in tall fescue have been evaluated for genetic variation. In addition, there have been many more that have been described in other C_3 grasses, both forage and grain crops, that should give insight into the genetic potentials of tall fescue. Much of the work has focused on ways to genetically improve adaptation, yield, forage quality, and seed production of tall fescue. Nearly all physiological traits studied have shown genetic variation. Disease and insect resistance, which are also genetically variable, are other important characters in tall fescue productivity.

A. Environmental Adaptation

Even though tall fescue is adapted to a wide range of environmental conditions, there have been numerous attempts to modify physiological processes associated with interactions between leaf growth and winterhardiness and between summer productivity and water use. The former has been researched most actively in Great Britain and Europe, the latter in the U.S.

British scientists recognized early that populations of tall fescue introduced from North Africa, primarily Tunisia and Morocco, had superior fall growth, but were not very winterhardy when grown under short days.[72,73] Plants of North African origin continued to have active leaf elongation at low temperatures, whereas those of European origin had markedly reduced leaf growth rates.[74] It is unclear at present whether the relationship between the two processes is a genuine cause-effect, i.e., active leaf growth interferes physiologically or is not compatible with development of winterhardiness. Conversely, the relationship may be a genetic linkage that can be broken through conventional or, more likely, through biotechnological procedures suggested elsewhere in this volume.

In any event, improving fall growth is a worthy objective for tall fescue, especially if winterhardiness is not impaired. Other research shows that soluble carbohydrates, mainly sucrose and fructan, are accumulated in the leaf blade tissues during cool fall weather.[75] Those carbohydrates contribute to forage quality of fall-saved herbage and will likely help the leaf blade resist freezing damage, senescence, and solute leakage. Thus, forage quality may be retained after freezing and animal performance may be enhanced on the forage grown under cool conditions.

In mid latitudes of the U.S., tall fescue exhibits good regrowth in late spring, but does not grow well during the warm, dry periods of July and August. Irrigation during summer increased herbage production,[76] but production was less during fall and the following spring, compared with plants that had not been summer irrigated. This caused some research groups to shift attention to C_4 species and alternate forages for summer production to allow tall fescue pastures to rest during summer, recognizing that fall recovery will be better.

Attempts to increase summer production by reducing water loss of tall fescue have not been successful. Horst and Nelson[77] screened tall fescue genotypes for leaf photosynthesis and transpiration during summer and found a strong positive relationship between the two gas exchange processes. This was attributed largely to stomatal conductance which influences both CO_2 and H_2O exchange with the atmosphere. Genotypes showed little, if any, variation in the degree of water stress required to close stomata. Improvement of water use efficiency by maintaining CO_2 uptake while decreasing H_2O loss has not been accomplished with conventional technology.

Buckner et al.[78] noted some tall fescue cultivars differed in leaf blade rolling during water stress periods. They were interested in maintaining succulence and suggested using minimum leaf roll as a selection factor. Inward rolling of the blade, however, decreases water loss because stomatal density is highest on the upper surface and rolling decreases the effective area of the leaf that intercepts radiation. If blades are unrolled during early morning and late afternoon for active photosynthesis, the water-conserving action of rolling during midday may give a long-term water use benefit. Plants that had minimum leaf roll also had fewer and smaller stomata, perhaps decreasing the water loss and retaining succulence. This strategy for improving water status and forage quality needs to be evaluated more extensively at the physiological level.

A major factor in adaptation is maturity or time of flowering, yet this character shows little genetic variation in Missouri.[79] Bean[80,81] in Great Britain has extensively studied the floral induction and initiation processes for tall fescue. The vernalization requirement is moderately strong in Missouri. Nearly 100% of the plants removed from the field on January 1 and placed in a long-day environment will initiate flowers, and the initiation process is quantitative with day length.[82] No explanation is available for why little variation exists in time of flowering in Missouri, yet that is an important cultural factor. Data for other species indicate that selection for later maturity may improve forage quality, but early maturity cultivars of tall fescue have highest seed yield in Missouri.[83,84] The maturity factor needs more genetic and physiological evaluation.

B. Improving Yield

Most breeders have improvement of forage yield as a primary objective. Since tall fescue has a moderately strong vernalization requirement, reproductive growth in spring is the major portion of the annual herbage production. Subsequent regrowths rarely produce elongated culms; thus, summer and fall yields are predominately leaf blades. It is generally recognized that light interception is a primary factor in yield enhancement.

In contrast with photosynthesis per unit leaf area, rate of leaf area production was positively related to forage yield of vegetative canopies.[23] This lead to an active genetic selection program for leaf growth rate. Gain from selection was still being made after five generations in both the low and high directions,[85] resulting in populations differing by more than twofold in leaf area expansion rate.[16] Over several experiments the diverse populations were similar or differed only slightly in photosynthetic rate, indicating that the leaf growth rate and subsequent leaf size was independent from photosynthetic rate per unit leaf area.

An indirect response to selection for rapid leaf elongation rate was reduced tiller production and a lower tiller density in established canopies. Thus, selection for rapid leaf growth led to a taller, more upright canopy that yielded more and allowed more light to penetrate into the canopy. The reduced tillering caused less competition on associated legumes, allowing improved compatibility between species.[46]

In subsequent studies we found that leaf elongation is not limited by carbohydrate supply, as content of WSC in the leaf-growth zone can exceed 40% of the dry weight of both fast and slow leaf growth genotypes.[38] This suggests that leaf growth is sink limited, i.e., leaf elongation and sucrose or carbohydrate use are regulated by some mechanism other than carbohydrate supply. Addition of GA_3 doubled the leaf elongation rate of both a decaploid and a hexaploid genotype of tall fescue, largely by increasing cell size.[43] Photosynthesis per unit leaf area was reduced by the GA_3 treatment, but to a lesser degree than the increase in leaf area expansion rate. We have not studied the carbohydrate economy of that system.

Addition of nitrogen has a profound effect on leaf elongation rate that is mediated almost entirely by increased cell production rate.[45] With both high and low nitrogen, the rate of cell growth and the final cell size is near equal.[44] Similarly, the respiration rates associated with the elongation process are similar for both slow and rapid leaf growth genotypes.[86] These experiments suggest that the genetic regulation of leaf growth rates is more associated with growth regulator activity or nitrogen metabolism than it is with carbohydrate supply in the leaf. It is becoming more clear that genetically increasing photosynthesis alone will not increase leaf growth rate or herbage yield.

VII. DEVELOPING AN IDEOTYPE

As a working physiologist-breeder team, we are concerned about the so-called "yield barrier" for forage grasses. At the same time, we are starting to understand some of the major factors that contribute to yield. Plants that are capable of surviving the environmental stresses and cultural practices that are imposed on forages will likely have a highly buffered yield potential. Even so, it is time to integrate knowledge among researchers in management, physiology, and breeding to determine factors associated with yield potential and their genetic control.

At the University of Missouri, we are attempting to understand the influence of major environmental factors and to develop populations of tall fescue that have rapid rates of leaf area expansion, an upright canopy, and high rates of net photosynthesis for infrequent cutting or rotational grazing. While that shape may not be most desirable for continuous grazing,[19,59] we feel that it is one approach to optimize solar energy interception and use, especially with the interest of our farmers in fall growth and winter grazing.

In conjunction with work on leaf area expansion, conventional polyploid crosses and recurrent selection are being used to develop strains of tall fescue with high CER. We are also investigating dark respiration of leaf blades to improve conservation of fixed carbon.[87,88] Ultimately, when canopy shape and size are manipulated to maximize sink activity and optimize light distribution in the canopy, we feel the next most logical way to further increase yield would be to adjust the carbon economy of the leaves to make more efficient use of the solar energy received.[50]

Next, we want to look at the stored energy level in the plants. Is it necessary to have large amounts of stored energy in the stem bases, or is there a "threshold level"? If we don't need large amounts of stored energy in the stem base, will carbohydrate stay in the herbage to improve quality, and ultimately animal performance? Or could leaving sufficient leaf area after cutting or grazing overcome even these limitations? This latter concern about stored energy will likely be controlled more by management systems than by breeding for specific plant types, but we know that selection for specific canopy types influences concentration of WSC.[59] We need to test our new genetic populations to determine if high CER or low dark respiration of leaves influences concentrations of WSC in the stem bases and leaf tissue.

Because of the need for fossil fuel energy for nitrogen fertilizer manufacture, we are interested in developing cultivars of tall fescue that are specifically adapted for growing with legumes such as red clover (*Trifolium pratense* L.), white clover (*T. repens* L.), or birdsfoot trefoil (*Lotus corniculatus* L.). Likely, tall fescue cultivars that are well adapted for mixtures will be more upright growing and have reduced tillering capacity to give less competition to the legume.[46]

Yield of grasses like tall fescue needs to be, and will be, improved significantly. Improved management will help in the short term. In the long term, improvement by breeding will allow higher yield plateaus to be achieved. Likely, physiology will play a key role in assisting breeders in developing the needed selection criteria. From a personal view, that physiological emphasis will need to be concentrated on improving light interception and carbon conservation in the canopy, and on increasing sink activity. While the physiology and interrelationships of these processes will be difficult to understand, use of the information for developing rapid, efficient selection criteria for breeders, or for developing feasible management systems, may be an even greater challenge. While physiological information by itself is valuable, its ultimate use through improvement in cultivars or cultural practices should prove to be invaluable.

VIII. PROBLEMS WITH CONVENTIONAL BREEDING METHODS FOR PHYSIOLOGICAL CHARACTERS

Sleper and Nelson (Chapter 9 of this book) discuss the use of haploids in a breeding program. Here we would like to address the issue more specifically in terms of genetic selection for physiological characters. A major consideration is that, to date, most of the physiological characters evaluated are regulated by additive gene action using a few (mineral concentration), a moderate number (dark respiration, disease resistance), or many loci (photosynthesis, leaf elongation rate, tillering, seed yield). This makes working with physiological traits in breeding programs difficult due to the large number of populations and plants per population that need to be evaluated.

Many breeding-physiology teams have addressed the numbers problem by placing the major emphasis on clonal parental selection. Clonal parents representing desirable agronomic phenotypes are selected from the characterized population, anticipating that the desired physiological character will be expressed. Crosses are made among selected clones, then generations are advanced and tested using traditional breeding methodology. Sometimes, the final populations are screened to determine if the desired trait from the parental population was retained.

For our genetic improvement program, however, we first evaluate genetic variation among individuals in a broad-based population,[49,87-89] then determine the gene action and heritability. Based on those aspects, a selection method is used that is appropriate for the gene action of the trait. Because traits are controlled by many genes, most often this involves some variation of recurrent selection.[16] We select in both the high and low directions so we can evaluate the physiological impact of the character, regardless of the hypothesis regarding its practical utility. The response to recurrent selection over generations allows evaluation of gene action.[85]

There is a need to evaluate large numbers of individuals within populations; it requires measurements on 24 to 40 plants per population for CER, leaf elongation, and tillering rate to get accurate assessment.[50] Further, there is reversion between generations due to segregation, environmental variation, and genetic recombination. For example, after four generations of recurrent selection for CER, the high population averaged 32.0 and the low population averaged 27.4 mg CO_2 dm^{-2} h^{-1}. Standard deviations were 4.6 and 4.8, respectively. The 10% selected from the fourth generation to intercross to produce the fifth generation averaged 37.3 and 20.8 mg CO_2 dm^{-2} h^{-1}, respectively. The fifth generation populations, however, averaged 33.8 and 23.7 mg CO_2 dm^{-2} h^{-1}, respectively, showing that progeny means were not as divergent as the selected parents.

As indicated earlier, our long-term objective is to genetically combine desirable parents selected from the low- and high-CER populations in all combinations with parents selected from the low and high leaf elongation populations. Because genotypes may differ in general and specific combining ability, we prefer to have at least four parents selected from each population. Using the selected parents and mating them in all combinations would require many crosses to be made. We would like to evaluate at least 40 progeny for CER, leaf growth, and high herbage yield for each generation. It is readily apparent why progress has been difficult when using this approach due to the large number of plants to be evaluated.

Based on our earlier work, we would like to combine as many as possible of the following characters into an improved cultivar.

1. High CER[49]
2. Rapid leaf area expansion rate[16]
3. Low dark respiration of mature leaves[90]
4. High forage quality[79]
5. High seed yield[83]
6. Favorable mineral ratio[91]

Genetic variability and heritability estimates have been determined for each character, and desirable plants and populations based on each characteristic have been identified. We have considered methods of simultaneous selection for combining characters. However, because of the complexity of inheritance and the large number of plants required, simultaneous selection procedures are expected to have limited success in combining many characters.

Similarly, if we chose to use tandem selection, the number of characters combined would be limited because of the time involved. Tandem selection involves selecting for one trait until it reaches a satisfactory level and then initiating selection on another trait until it is at a satisfactory level, etc. This procedure would be relatively inefficient in combining several physiological characters.

Likewise, independent culling poses a challenge because of the large number of progeny required. For example, if we chose to simultaneously select for both high CER and high leaf area expansion rate with a selection intensity of 10% for each character, we would be selecting only the upper 1% of the population. This assumes that these two characters are inherited independently, for which we have some evidence.[16,43] As one increases the number of characters selected for simultaneously using independent culling, one quickly concludes that it is difficult, if not impossible, because of the large number of progeny that need to be evaluated.

IX. POTENTIAL USE OF HAPLOIDS AND DOUBLED HAPLOIDS TO COMBINE PHYSIOLOGICAL CHARACTERS

It is recognized that genetic improvement of grasses such as tall fescue is very slow using conventional methods.[92] Therefore, advantages of using haploids and their doubled haploids would be more rapid identification of haploid plants with desired physiological characters and the ability to fix desirable alleles for physiological traits in homozygous lines after chromosome doubling.

Haploids of tall fescue have been developed by culturing anthers[93] and from synergid twins using polyembryonic seedlings.[70,79] These haploids tend to be stable and can be maintained by cloning via vegetative tillers (see

Chapters 4 to 7 and 9 of this book). Further, the haploid plants obtained from heterozygous tall fescue display plant-to-plant phenotypic and physiological differences in characters such as growth rate,[94] mineral ratios,[71] and CER.[70] Once populations of haploids have been developed, they can be cloned via tillers, and selection can occur for a range of characters. Then, chromosome numbers of the best individuals can be doubled (see Chapter 6 of this book) to develop fertile homozygous doubled haploids with the characteristics identified in the selected "parent" haploid plant.

After developing fertile homozygous doubled haploid plants with high CER and others with high leaf area expansion, F_1 hybrids could be made. The hybrids would then be tested as experimental hybrid cultivars which would likely possess maximum expression for both traits and possibly heterosis. Heterosis has been reported in tall fescue using conventional plant breeding procedures.[95] We have also noted heterosis for biomass yield when tall fescue plants selected for high CER were crossed conventionally with plants selected for high leaf area expansion rate.[50] Hybrids of doubled haploids should result in even more pronounced heterosis. One could easily combine three traits by making three-way crosses and four traits by making double crosses. Through the use of gamete-derived haploids, selection among the haploids, and derivation of doubled haploids from selected haploids, it is theoretically possible to have multiple physiological traits fixed in high-yielding hybrids and/or synthetic cultivars.

X. SUMMARY

Tall fescue is grown in the east-central U.S. primarily as a turf species and as a pasture species for beef cattle. It is particularly adapted to deferred grazing in autumn and winter because leafy regrowth continues through late autumn. We are attempting to improve productivity by understanding physiological characteristics related to herbage yield. Once identified, these traits can serve as selection criteria for the use of genetic improvement. Our research shows that cultivars developed for infrequent grazing or cutting should have rapid leaf area expansion rates, upright leaves, active photosynthetic rate, low dark respiration rate of leaf blades, and ability to store adequate quantities of WSC in the stem bases. Cultivars developed for frequent grazing should have rapid tiller formation, which will provide residual leaf area after grazing, to provide energy for regrowth via photosynthesis because concentration of stored WSC will be low.

The rapid leaf growth types will be most compatible with leguminous species because of the more open canopy. Contrasting populations of tall fescue that have been developed for high and low photosynthetic rate, high and low leaf elongation rate, and other physiological characters are presently being crossed and progeny evaluated to aid our understanding of the interactions of these physiological traits with each other and on herbage yield.

Conventional breeding techniques require large populations to handle multiple characters.

Haploid plants derived from immature gametes of heterozygous tall fescue plants should include a wide range of genotypes for various physiological characters. The haploid plants can be cloned via tillers and screened for a number of physiological characters. Doubling chromosome numbers in the most superior haploids should result in rapid development of fertile homozygous lines for selected characters. The doubled haploids should be useful in development of hybrids or synthetic cultivars. Use of these biotechnological approaches in combination with conventional wisdom for genetic improvement of physiological processes in tall fescue should result in more rapid improvement of this important species.

REFERENCES

1. **Hartley, W.,** The global distribution of tribes of the Gramineae in relation to historical and environmental factors, *Aust. J. Agric. Res.,* 1, 355, 1950.
2. **Leasure, J. K.,** The growth pattern of mixtures of orchardgrass and tall fescue with ladino clover in relation to temperature, *Proc. Assoc. South. Agric. Work.,* 49, 177, 1952.
3. **Rountree, B. H., Matches, A. G., and Martz, F. A.,** Season too long for your grass pasture?, *Crops Soils Mag.,* 26 (7), 7, 1974.
4. **Gilbert, W. B. and Chamblee, D. S.,** Effect of depth of water table on yield of ladino clover, orchardgrass, and tall fescue, *Agron. J.,* 57, 502, 1965.
5. **Badoux, S.,** Breeding tall fescue (*Festuca arundinacea* Schreb.) and cocksfoot (*Dactylis glomerata* L.) for seedling establishment under conditions of competition, in *Proc. XIII Int. Grassland Congr.,* Sectional Papers 1, 1977, 443.
6. **Pederson, G. A.,** White clover seed germination in agar containing tall fescue leaf extracts, *Crop Sci.,* 26, 1248, 1986.
7. **Peters, E. J. and Mohammed Zam, A. H. B.,** Allelopathic effects of tall fescue genotypes, *Agron. J.,* 73, 56, 1981.
8. **Luu, K. T., Matches, A. G., Nelson, C. J., Peters, E. J., and Garner, G. B.,** Characterization of inhibitory substances of tall fescue on birdsfoot trefoil, *Crop Sci.,* 29, 407, 1989.
9. **Bush, L. P. and Burrus, P. B.,** Tall fescue quality and agronomic performance as affected by the endophyte, *J. Prod. Agric.,* 1, 55, 1988.
10. **Clay, K., Hardy, T. N., and Hammond, A. M.,** Fungal endophytes of grasses and their effects on an insect herbivore, *Oecologia,* 66, 1, 1985.
11. **West, C. P., Izekor, E., Oosterhuis, D. M., and Robbins, R. T.,** The effect of *Acremonium coenophialum* on the growth and nemotode infestation of tall fescue, *Plant Soil,* 112, 3, 1988.
12. **Nelson, C. J., Asay, K. H., and Sleper, D. A.,** Mechanisms of canopy development of tall fescue genotypes, *Crop Sci.,* 17, 449, 1977.
13. **Brougham, R. W.,** Effect of intensity of defoliation on regrowth of pasture, *Aust. J. Agric. Res.,* 7, 377, 1956.

14. **Brougham, R. W.,** The effects of frequent hard grazings at different times of the year on the productivity and species yields of a grass-clover pasture, *N. Z. J. Agric. Res.*, 3, 125, 1960.
15. **Hunt, L. A. and Brougham, R. W.,** Some changes in the structure of a perennial ryegrass sward frequently, but leniently defoliated during the summer, *N. Z. J. Agric. Res.*, 10, 397, 1967.
16. **Nelson, C. J. and Sleper, D. A.,** Using leaf area expansion rate to improve yield of tall fescue, in *Proc. XIV Int. Grassland Congr.*, Smith, J. A. and Hays, V. W., Eds., Westview Press, Boulder, CO, 1983, 413.
17. **Hart, R. H., Carlson, G. E., and McCloud, D. E.,** Cumulative effects of cutting management on forage yields and tiller densities of tall fescue and orchardgrass, *Agron. J.*, 63, 895, 1971.
18. **Nelson, C. J. and Coutts, J. H.,** *Physiological Responses of Tall Fescue to Nitrogen Fertilization,* Abstract, American Society of Agronomy, Madison, WI, 1979, 106.
19. **Rhodes, I.,** Relationship between canopy structure and productivity in herbage grasses and its implications for plant breeding, *Herb. Abstr.*, 43, 129, 1973.
20. **Rhodes, I.,** Yield, canopy structure and light interception of two ryegrass varieties in mixed culture and monoculture, *J. Br. Grassl. Soc.*, 24, 123, 1969.
21. **Rhodes, I.,** The relationship between productivity and some components of canopy structure in ryegrass (*Lolium* spp.). II. Yield, canopy structure and light interception, *J. Agric. Sci.*, 77, 283, 1971.
22. **Rhodes, I. and Mee, S. S.,** Changes in dry matter yield associated with selection for canopy characters in ryegrass, *Grass Forage Sci.*, 35, 35, 1980.
23. **Horst, G. L., Nelson, C. J., and Asay, K. H.,** Relationship of leaf elongation to forage yield of tall fescue genotypes, *Crop Sci.*, 18, 715, 1978.
24. **Cooper, J. P. and Edwards, K. J. R.,** The genetic control of leaf development in *Lolium*. I. Assessment of genetic variation, *Heredity,* 16, 63, 1961.
25. **Cooper, J. P. and Edwards, K. J. R.,** The genetic control of leaf development in *Lolium*. II. Response to selection, *Heredity,* 18, 307, 1963.
26. **Davidson, D. J. and Nelson, C. J.,** *Growth and Development of Tall Fescue in Relation to its Plastochron Index,* Abstract, American Society of Agronomy, Madison, WI, 1979, 86.
27. **Booysen, P. and Nelson, C. J.,** Leaf area and carbohydrate reserves in regrowth of tall fescue, *Crop Sci.*, 15, 262, 1975.
28. **Leafe, E. L. and Parsons, A. J.,** Physiology of growth of a grazed sward, in *Proc. XIV Int. Grassland Congr.*, Smith, J. A. and Hays, V. W., Eds., Westview Press, Boulder, CO, 1983, 403.
29. **Parsons, A. J. and Leafe, E. L.,** Photosynthesis and carbon balance of a grazed sward, in *Plant Physiology and Herbage Production,* Wright, C. E., Ed., British Grassland Society, Hurley, 1981, 69.
30. **Zarrough, K. M., Nelson, C. J., and Sleper, D. A.,** Interrelationships between rates of leaf appearance and tillering in selected tall fescue populations, *Crop Sci,* 24, 565, 1984.
31. **Nelson, C. J., Asay, K. H., and Horst, G. L.,** Relationship of leaf photosynthesis to forage yield of tall fescue, *Crop Sci.*, 15, 476, 1975.
32. **Wilhelm, W. W. and Nelson, C. J.,** Growth analysis of tall fescue genotypes differing in yield and leaf photosynthesis, *Crop Sci.*, 18, 951, 1978.
33. **Sleper, D. A., Nelson, C. J., and Asay, K. H.,** Diallel and path coefficient analysis of tall fescue (*Festuca arundinacea*) regrowth under controlled conditions, *Can. J. Genet. Cytol.*, 19, 557, 1977.
34. **Volenec, J. J. and Nelson, C. J.,** Cell dynamics in leaf meristems of contrasting tall fescue genotypes, *Crop Sci.*, 21, 381, 1981.

35. **Wilson, D.,** Leaf growth, stomatal diffusion resistances, and photosynthesis during droughting of *Lolium perenne* populations selected for contrasting stomatal length and frequency, *Ann. Appl. Biol.*, 79, 67, 1975.
36. **Volenec, J. J. and Nelson, C. J.,** Diurnal leaf elongation of contrasting tall fescue genotypes, *Crop Sci.*, 22, 531, 1982.
37. **Schnyder, H. and Nelson, C. J.,** Diurnal growth of tall fescue leaf blades. I. Spatial distribution of growth, deposition of water and assimilate transport in the elongation zone, *Plant Physiol.*, 86, 1070, 1988.
38. **Volenec, J. J. and Nelson, C. J.,** Carbohydrate metabolism in leaf meristems of tall fescue. I. Relationship to genetically altered leaf elongation rates, *Plant Physiol.*, 74, 590, 1984.
39. **Spollen, W. G. and Nelson, C. J.,** Characterization of fructan from mature leaf blades and elongation zones of developing leaf blades of wheat, tall fescue, and timothy, *Plant Physiol.*, 88, 1349, 1988.
40. **Schnyder, H. and Nelson, C. J.,** Growth rates and carbohydrate fluxes within the elongation zone of tall fescue leaf blades, *Plant Physiol.*, 85, 548, 1987.
41. **Moser, L. E., Volenec, J. J., and Nelson, C. J.,** Respiration, carbohydrate content and leaf growth of tall fescue, *Crop Sci.*, 22, 781, 1982.
42. **Volenec, J. J. and Nelson, C. J.,** Carbohydrate metabolism in leaf meristems of tall fescue. II. Relationship to leaf elongation rates modified by nitrogen fertilization, *Plant Physiol.*, 74, 595, 1984.
43. **Poskuta, J. W., Nelson, C. J., Coutts, J. H., and Vassey, T. L.,** Effect of gibberellic acid on photosynthesis and leaf characteristics of tall fescue, *Curr. Top. Plant Biochem. Physiol.*, 3, 174, 1984.
44. **MacAdam, J. W., Volenec, J. J., and Nelson, C. J.,** Effects of nitrogen on mesophyll cell division and epidermal cell elongation in tall fescue leaf blades, *Plant Physiol.*, 89, 549, 1989.
45. **Volenec, J. J. and Nelson, C. J.,** Responses of tall fescue leaf meristems to N fertilization and harvest frequency, *Crop Sci.*, 23, 720, 1983.
46. **Zarrough, K. M., Nelson, C. J., and Coutts, J. H.,** Relationship between tillering and forage yield of tall fescue. I. Yield, *Crop Sci.*, 23, 333, 1983.
47. **Zarrough, K. M., Nelson, C. J., and Coutts, J. H.,** Relationship between tillering and forage yield of tall fescue. II. Pattern of tillering, *Crop Sci.*, 23, 338, 1983.
48. **Zarrough, K. M. and Nelson, C. J.,** Unpublished data, 1982.
49. **Asay, K. H., Nelson, C. J., and Horst, G. L.,** Genetic variability for net photosynthesis in tall fescue, *Crop Sci.*, 14, 571, 1974.
50. **Nelson, C. J.,** Genetic association between photosynthetic characteristics and yield: Review of the evidence, *Plant Physiol. Biochem.*, 26, 543, 1988.
51. **Sheehy, J. E. and Cooper, J. P.,** Light inteception, photosynthetic activity, and crop growth rate in canopies of six temperate forage grasses, *J. Appl. Ecol.*, 10, 239, 1973.
52. **Hoveland, C. S., Foutch, H. W., and Buchanan, G. A.,** Response of *Phalaris* genotypes and other cool-season grasses to temperature, *Agron. J.*, 66, 686, 1974.
53. **Nelson, C. J., Asay, K. H., and Horst, G. L.,** Relationship of leaf photosynthesis of forage yield of tall fescue, *Crop Sci.*, 15, 476, 1975.
54. **Woledge, J.,** The effect of shading during vegetative and reproductive growth on the photosynthetic capacity of leaves in a grass sward, *Ann. Bot.*, 42, 1085, 1978.
55. **Wilson, D. and Cooper, J. P.,** Effect of selection for mesophyll cell size on growth and assimilation in *Lolium perenne* L., *New Phytol.*, 69, 233, 1970.
56. **Byrne, M. C., Nelson, C. J., and Randall, D. D.,** Ploidy effects on anatomy and gas exchange of tall fescue leaves, *Plant Physiol.*, 69, 891, 1981.
57. **Nelson, C. J. and Sleper, D. A.,** Unpublished data, 1986.
58. **May, L. H.,** The utilization of carbohydrate reserves in pasture plants after defoliation, *Herb. Abstr.*, 30, 239, 1960.

59. **Zarrough, K. M. and Nelson, C. J.,** Regrowth of genotypes of tall fescue differing in yield per tiller, *Crop Sci.,* 20, 540, 1980.
60. **Volenec, J. J.,** Nonstructural carbohydrates in stem base components of tall fescue during regrowth, *Crop Sci.,* 26, 122, 1986.
61. **Smith, D.,** Growth and development of timothy tillers as influenced by level of carbohydrate reserves and leaf area, *Ann. Bot.,* 38, 595, 1974.
62. **Ward, C. Y. and Blaser, R. E.,** Carbohydrate reserves and leaf area in regrowth of orchardgrass, *Crop Sci.,* 1, 366, 1961.
63. **Davidson, D. J., Nelson, C. J., and Randall, D. D.,** *Nitrogen Effects on Photosynthesis and Senescence of Tall Fescue Leaves,* Abstract, American Society of Agronomy, Madison, WI, 1980, 78.
64. **Randall, D. D., Nelson, C. J., Sleper, D. A., Miles, C. D., Crane, C. F., Krueger, R. W., Wong, J. H. H., and Poskuta, J. W.,** Photosynthesis in allopolyploid *Festuca,* in *Nitrogen Fixation and CO_2 Metabolism,* Ludden, P. W. and Burris, J. E., Eds., Elsevier, New York, 1985, 409.
65. **Randall, D. D., Nelson, C. J., and Asay, K. H.,** Ribulose bisphosphate carboxylase: altered genetic expression in tall fescue, *Plant Physiol.,* 59, 38, 1977.
66. **Joseph, M. C., Randall, D. D., and Nelson, C. J.,** Photosynthesis in polyploid tall fescue. II. Photosynthesis and ribulose-1,5-bisphosphate carboxylase of tall fescue, *Plant Physiol.,* 68, 894, 1981.
67. **Poskuta, J. W. and Nelson, C. J.,** Oxygen and radiation effects on CO_2 exchange in light and in darkness of decaploid and hexaploid tall fescue (*Festuca arundinacea* Schreb.), *Acta Soc. Bot. Pol.,* 55, 35, 1986.
68. **Krueger, R. W. and Miles, C. D.,** Photosynthesis in fescue. III. Rates of electron transport in a polyploid series of tall fescue plants, *Plant Pysiol.,* 68, 1110, 1981.
69. **Allard, G., Nelson, C. J., and Pallardy, S. G.,** Shading effects on assimilation rate and internal CO_2 of tall fescue leaves, *Curr. Top. Plant Biochem. Physiol.,* 6, 166, 1987.
70. **Poskuta, J. W., Nelson, C. J., and Sleper, D. A.,** CO_2 exchange, leaf area and herbage yield of tetraploid and polyhaploid *Festuca* spp., *Photosynthetica,* 22, 341, 1988.
71. **Kasperbauer, M. J., Karlen, D. L., and Burton, H. R.,** Ploidy effects on protein, in vitro dry matter disappearance, and potassium/(calcium + magnesium) equivalent ratio in tall fescue forage, *Crop Sci.,* 27, 1081, 1987.
72. **MacCall, D. and Cooper, J. P.,** Climatic variation in forage grasses. III. Seasonal changes in growth and assimilation in climatic races of *Lolium, Dactylis,* and *Festuca, J. Appl. Ecol.,* 4, 113, 1967.
73. **Robson, M. J. and Jewiss, O. R.,** A comparison of British and North African varieties of tall fescue (*Festuca arundinacea*). III. Effects of light, temperature, and daylength on relative growth rate and its components, *J. Appl. Ecol.,* 5, 191, 1968.
74. **Nelson, C. J., Treharne, K. J., and Cooper, J. P.,** Influence of temperature on leaf growth of diverse populations of tall fescue, *Crop Sci.,* 18, 217, 1978.
75. **Brown, R. H. and Blaser, R. E.,** Relationships between reserve carbohydrate accumulation and growth rate in orchardgrass and tall fescue, *Crop Sci.,* 5, 577, 1965.
76. **Horst, G. L. and Nelson, C. J.,** Compensatory growth of tall fescue following drought, *Agron. J.,* 71, 559, 1979.
77. **Horst, G. L. and Nelson, C. J.,** Unpublished data, 1975.
78. **Buckner, R. C., Bush, L. P., and Burrus, P. B.,** Succulence as a selection criterion for improved forage quality in *Lolium-Festuca* hybrids, *Crop Sci.,* 19, 93, 1979.
79. **Sleper, D. A.,** Breeding tall fescue, *Plant Breed. Rev.,* 3, 313, 1985.
80. **Bean, E. W.,** Environmental and genetic effects upon reproductive growth in tall fescue (*Festuca arundinacea* Schreb.), *J. Agric. Sci.,* 72, 341, 1969.
81. **Bean, E. W.,** Clonal evaluation for increased seed production in two species of forage grasses, *Festuca arundinacea* Schreb. and *Phleum pratense* L., *Euphytica,* 21, 377, 1972.

82. **Nelson, C. J. and Sleper, D. A.,** Unpublished data, 1980.
83. **Nguyen, H. T. and Sleper, D. A.,** Diallel analysis of seed yield and reproductive characters in two populations of tall fescue, *Z. Pflanzenzuecht.*, 94, 111, 1985.
84. **Nguyen, H. T. and Sleper, D. A.,** Genetic variability of seed yield and reproductive characters in tall fescue, *Crop Sci.*, 23, 621, 1983.
85. **Reeder, L. R., Sleper, D. A., and Nelson, C. J.,** Response to selection for leaf area expansion rate of tall fescue, *Crop Sci.*, 24, 97, 1984.
86. **Pearen, J. R. and Nelson, C. J.,** *Respiratory Efficiency of Tall Fescue Leaf Growth,* Abstract, American Society of Agronomy, Madison, WI, 1988, 115.
87. **Jones, R. J. and Nelson, C. J.,** Respiration and concentration of water soluble carbohydrate in plant parts of contrasting tall fescue genotypes, *Crop Sci.*, 19, 367, 1979.
88. **Volenec, J. J., Nguyen, H. T., Nelson, C. J., and Sleper, D. A.,** Potential for genetically modifying dark respiration of tall fescue leaves, *Crop Sci.*, 24, 938, 1984.
89. **Nelson, C. J., Asay, K. H., and Patton, L. D.,** Photosynthetic responses of tall fescue to selection for longevity below the CO_2 compensation point, *Crop Sci.*, 15, 629, 1975.
90. **Volenec, J. J., Nelson, C. J., and Sleper, D. A.,** Influence of temperature on leaf dark respiration of diverse tall fescue genotypes, *Crop Sci.*, 24, 907, 1984.
91. **Sleper, D. A., Garner, G. B., Nelson, C. J., and Sebaugh, J. L.,** Mineral concentration of tall fescue genotypes grown under controlled conditions, *Agron. J.*, 72, 720, 1980.
92. **Hanson, A. A.,** The future of tall fescue, in *Tall Fescue, (Monogr. 20),* Buckner, R. C. and Bush, L. P., Eds., American Society of Agronomy, Madison, WI, 1979, 341.
93. **Kasperbauer, M. J., Buckner, R. C., and Springer, W. D.,** Haploid plants by anther-panicle culture of tall fescue, *Crop Sci.*, 20, 103, 1980.
94. **Kasperbauer, M. J. and Eizenga, G. C.,** Tall fescue doubled haploids via tissue culture and plant regeneration, *Crop Sci.*, 25, 1091, 1985.
95. **Moutray, J. B., Jr. and Frakes, R. V.,** Effects of genetic diversity on heterosis in tall fescue, *Crop Sci.*, 13, 1, 1973.

Chapter 9

BREEDING AND GENETICS: POTENTIAL USE OF HAPLOIDS AND DOUBLED HAPLOIDS

D. A. Sleper and C. J. Nelson

TABLE OF CONTENTS

I. INTRODUCTION

Tall fescue (*Festuca arundinacea* Schreb.) is a grass species that is used widely for forage and turf. It is estimated that there are 14 million ha of tall fescue used for forage in the U.S.[1] During the past 15 years, tall fescue has also received eminent status for its use as a turf species. Its use as a turf and forage species in the U.S. is confined largely to the transition zone. The transition zone is defined as that area in the U.S. that exists between the successful cultivation of cool- and warm-season grasses. This area is bordered on the western edge by eastern Oklahoma, Kansas, and Nebraska and extends to the east coast. The southern boundary includes northern Texas, Louisiana, Alabama, and approximately the northern half of Georgia. The northern boundary includes southern Iowa and extends to the east coast at a similar latitude.

"Kentucky 31" was one of the first tall fescue cultivars released in the U.S. It was first evaluated in 1931 and released in 1943. Kentucky 31 was quickly recognized for its adaptation to a wide range of soils and its ability to grow in the east-central part of the U.S., where temperatures are often too warm for orchardgrass (*Dactylis glomerata* L.) and Kentucky bluegrass (*Poa pratensis* L.). Kentucky 31 was also persistent and dependable, especially with intensive grazing by farm livestock. It is adapted to a wide range of soils, is persistent, and can be closely grazed or allowed to accumulate for winter grazing (stockpiled). Expanded livestock production in the transition zone would not have been possible without the use of tall fescue, primarily the Kentucky 31 cultivar.

Tall fescue breeders have developed many improved cultivars that are successfully cultivated in the transition zone. To a large extent, emphasis is on breeding for improved animal performance. In the transition zone of the U.S., there is a primary concentration of beef cows. The major emphasis is to produce and grow calves to weights of about 230 kg. Calves are then sold to other farmers for further feeding to weights of about 500 kg for marketing as meat. About 80 to 85% of the feed units for the cow-calf unit are derived directly from grazed or harvested forage, and in much of the lower corn belt (transition zone) that forage is tall fescue.[2] Over 50% of the pastures of tall fescue exist as a monoculture, with many receiving some fertilizer nitrogen. There is a steady increase in improvement of tall fescue pastures by addition of legumes such as red clover (*Trifolium pratense* L.), white clover (*T. repens* L.), Lespedeza (*Lespedeza striata* Hook and Arn.) and, to a lesser extent, birdsfoot trefoil (*Lotus corniculatus* L.). Both animal performance and seasonality of production are improved by adding a legume. Today, adding a legume to tall fescue pastures known to be infected with the endophytic fungus (*Acremonium coenophialum* Morgan-Jones & Gams) is a management recommendation to improve animal performance.

The most recent advance in breeding tall fescue has been the release of endophyte (*A. coenophialum*)-free cultivars. With 14 million ha of tall fescue which is estimated to be highly infected with the endophytic fungus, the potential losses in dollars and in animal gains is tremendous. The use of these new endophyte-free cultivars has the potential to save producers millions of dollars. As an example, the endophytic fungus is prevalent at a significant level in 90% of Missouri pastures. This level of infection can potentially reduce animal gains by 22.7 kg per growing animal. The Missouri beef cow herd of 2.4 million head is expected to produce 1.92 million calves annually. Backgrounded steers and heifers represent another 860,000 animals. The calculated loss at $1.10/kg in Missouri alone is $63 million annually to the economy of the state (2.8 million calves and backgrounded steers and heifers × 22.7 kg × $1.10/kg × 90% = $63 million). Similar analyses conducted for other states in the tall fescue growing region add up to huge dollar losses annually in the nation. Data has shown that at least a 30% improvement in animal average daily gains can be realized by having animals graze endophyte-free cultivars rather than infected cultivars.[3]

Conventional tall fescue breeding is an endeavor that depends heavily on other scientific disciplines. These include plant pathology, plant physiology, animal science, cytogenetics, genetics, statistics, and entomology, just to name a few. Because tall fescue forage cultivars will likely be used in producing animals or animal products, most tall fescue breeders cooperate with animal scientists. This is particularly true in the final evaluation phases of cultivar development where experimental synthetics are evaluated for animal performance prior to release.

The inclusion of haploids and other biotechnologically derived plants in a tall fescue breeding program has the potential to serve as yet another tool in the development of new and improved cultivars. Before one can appreciate the potential usefulness of haploids in the breeding of tall fescue, an understanding of what has been and is presently state of the art in breeding improved cultivars is necessary.

II. PHYLOGENETIC RELATIONSHIPS

A. Geographical Distribution

The native distribution of tall fescue includes the areas on both the north and south sides of the Mediterranean Sea. It extends over much of the European continent and North Africa. It is found in the Tunisian area and in west and central Asia and Siberia.[4,5] Hexaploids distributed over this area are included in distinguishable ecogeographic races, and poor chromosome pairing relationships are often observed in F_1 progeny resulting from crosses between geographical races. Evans et al.[6] and Hunt and Sleper[7] took American accessions received from Europe and crossed these to accessions from Tunisia to produce F_1 hybrids. Meiotic irregularities occurred, including univalents and multivalents. Seed production on these hybrids was considerably less than that observed for hybrids made from hexaploid parents crossed from similar geographical regions. There is evidence to suggest that repeated backcrossing may improve upon the ability to transfer genes between geographic races.

Tall fescue today is cultivated in pastures far from its sources of origin. Tall fescue has successfully been introduced in North and South America, New Zealand, Australia, in East and South Africa, Japan, and more recently, mainland China.

B. Genomic Relationships

Cultivated tall fescue in the U.S. consists largely of hexaploid ($2n = 6x = 42$), *Festuca arundinacea*.[8] In Europe, both the hexaploid and the diploid ($2n = 2x = 14$), *F. pratensis* Huds., are widely cultivated. The *Festuca* species of interest to plant breeders and geneticists ranges in ploidy level from 2x to 10x (Table 1). The allopolyploid series[8-11] is founded on a basic chromosome number of $x = 7$. The allopolyploid nature of the *Festuca* species presents a challenge to the breeder because the polyploid composition has a complicating influence on phenotypic expression and genetic analysis. The diploid species *F. pratensis* is phenotypically quite similar to the hexaploid *F. arundinacea* var. genuina Schreb. The hexaploid is hypothesized to have one genome from *F. pratensis* and two related genomes from the tetraploid *F. arundinacea* var. glaucescens Boiss. In addition, plant breeders have crossed the hexaploid tall fescue to both annual (*Lolium multiflorum* L.) and perennial ryegrass (*L. perenne* L.).[12] The hexaploid tall fescue cultivar "Kenhy" was developed as the result of introducing genes and cytoplasm from annual

TABLE 1
Proposed Genomic Formula of *Festuca* Species

Species	Somatic number (2n)	Proposed genomic formula
F. pratensis Huds	14	PP
F. arundinacea var. glaucescens Boiss	28	$G_1G_1G_2G_2$
F. mairei St. Yves	28	$M_1M_1M_2M_2$
F. arundinacea var. genuina Schreb.	42	$PPG_1G_1G_2G_2$
F. arundinacea var. atlantigena St. Yves	56	$G_1G_1G_2G_2M_1M_1M_2M_2$
F. arundinacea var. letourneuxiana and cirtensis	70	$QQG_1G_1G_2G_2M_1M_1M_2M_2$

From Sleper, D. A., *J. Plant Breed. Rev.*, 3, 313, 1985. With permission.

ryegrass. Much more research needs to be conducted on the phylogenetic relationships among tall fescue and its relatives because this association is incompletely understood.

III. CONSIDERATIONS IN BREEDING

A. Types of Cultivars Grown Commercially

Commercial propagation of tall fescue is by seed. A large tall fescue seed marketing industry has developed in largely two parts of the nation. The largest portion of certified seed of improved cultivars is produced in the Willamette valley in the state of Oregon. A considerable amount of seed is also produced in the central part of the U.S., with the states of Missouri, Kentucky, and Tennessee producing the largest amounts. Most of the seed produced in the central part of the nation is marketed as uncertified seed.

Plant breeders and other scientists often propagate tall fescue vegetatively. However, it is not practical to propagate tall fescue vegetatively on a commercial scale. Vegetative propagation is possible because tall fescue produces numerous tillers. Vegetative propagation is of considerable use to plant breeders and other scientists, because it represents a rather rapid and inexpensive means of increasing valuable strains. For example, clonal replication represents an excellent means of establishing a polycross block for the production of breeders seed. Another advantage of vegetative propagation is that it allows researchers to conduct experiments with virtually identical genetic material. Having similar genetic material through vegetative cloning is valuable if one wants to keep genetic variation to a minimum and study the effects of certain environmental treatments.

Tall fescue cultivars are marketed as synthetics. A synthetic cultivar is defined as one in which its parents were selected for good general combining ability and mated *inter se* under conditions of random mating.[13] The newly developed synthetic cultivar is maintained through open pollination. The first generation is known as breeders seed and is referred to as the syn-1 generation. It takes several generations of seed increase to have sufficient seed available to satisfy the producers' demand. Producers usually obtain syn-4 or later generation seed. Because commercial seed is likely to be syn-4 or later, it is important that the plant breeder know the potential loss of performance due to inbreeding depression. It is recommended that variety trials not be initiated with syn-1 seed because maximum heterosis is expected for this generation. Most plant breeders will use syn-2 or later generation seed for inclusion into performance trials.

To minimize the impact of inbreeding depression by maintaining a broad genetic base, breeders often will produce synthetic cultivars with ten or more parents. Missouri 96, for example, has 13 parents.[14] However, a recently released cultivar, ''Martin'', has only two parents (unpublished). Martin is characterized by having superior performance over a wide range of environments. It is not clear how a two-clone synthetic which has been advanced for several generations can maintain a high level of performance without any undue inbreeding depression. It appears that more study is needed to accurately characterize the effects of inbreeding in tall fescue.

B. Breeding Objectives for Cultivar Development

Important objectives in the breeding of improved tall fescue cultivars includes high forage and seed yield, high herbage quality, and resistance to insects and diseases. Tall fescue is recognized as being a persistent perennial forage grass. Breeders located in northern regions of the U.S. may have as an objective winterhardiness. Breeders in the southern part of the U.S. are concerned with persistence under heat and drought stress during summer months. Throughout the central portion of the tall fescue belt, little emphasis is placed on persistence.

Heritability for herbage yields is recognized as being low for tall fescue and other forage grasses.[15] In most instances, some form of progeny testing is necessary to isolate superior genotypes capable of transmitting high herbage yield. However, testing of new strains in swards has not necessarily led to new cultivars with improved herbage yield. The difficulties in development of high herbage-yielding cultivars cannot solely be attributed to low heritability and inadequate progeny testing. Experimental strains of tall fescue are often tested under conditions that may not be found in producers' fields. For example, tall fescue is grown commercially under widely varying environmental conditions. Tall fescue may be grown on productive and rather unproductive soils, used as a pasture species in some instances and as a hay crop in others. The species may be grown with or without a legume. These varying environmental factors present a challenge in cultivar development to the plant breeder. In spite of these difficulties, breeding for improved herbage yield is important, particularly when emphasis is placed on high yield of digestible nutrients as contrasted with high herbage yield per hectare.

1. Herbage Yield

Increased emphases has been placed recently on selecting for components of yield to improve vegetative herbage yield.[8] Vegetative herbage yield in tall fescue has been reported to be a function of tillers per unit area of land and yield per tiller.[16] Research has also shown that number of tillers per plant was more important than yield per tiller in determining vegetative herbage yield per plant of tall fescue grown under conditions of low competition. Under well-developed swards, this would not be expected because tiller density becomes stabilized.[17,18] Jones et al.[19] selected tall fescue genotypes for differential herbage yield per tiller and found that these selections maintained their relative yield per tiller differences under sward conditions. Consistent with earlier results,[16,20] these genotypes displayed a negative relationship between yield per tiller and number of tillers per plant under field conditions.[17,18]

Another component of herbage yield in tall fescue is leaf area expansion rate (LAER).[21] High LAER is highly correlated with postreproductive herbage yield. The value of LAER (square millimeters per day) is estimated by taking the product of leaf expansion rate in millimeters per day and leaf width. The narrow sense heritability of LAER is less that 30% and is largely additive.[20] Again, as in herbage yield per se, the heritability is low. This heritability is low because leaf growth parameters are highly influenced by the environment.

Five cycles for high and low LAER have been completed with the objective of improving postreproductive herbage yield. A breeding procedure called recurrent restricted phenotypic selection[22,23] was used to select for high and low LAER because it minimizes the influence of the environment as contrasted to many other recurrent selection procedures.

After five cycles of selection, the mean increase in selecting for high LAER was greater than 10% per cycle; a mean decrease of more than 8% per cycle was observed when selecting for low LAER (Table 2).[24] A 4-year field study under sward conditions revealed that the high LAER had highest postreproductive herbage yields of all entries under infrequent defoliation, while the low LAER had highest postreproductive yields under frequent defoliation (unpublished). This differential response to defoliation indicates that these new experimental strains will respond differently to management practices.

2. Seed Yield

Selection for improved seed yield is of paramount importance. High-seed-yielding cultivars of tall fescue are attractive to those who produce cultivars only for seed. Therefore, companies marketing low-seed-yielding cultivars have to contract for higher seed prices, which makes it more costly to growers wanting to use the new cultivar as a pasture or hay crop.

Breeding for improved seed yield presents a special challenge to the tall fescue breeder because much of the certified seed is produced in the north-

TABLE 2
Response to Divergent Selection for Leaf Area Expansion Rate (LAER) on Tall Fescue

	LAER (mm^2/d)	
Entry	**1980**	**1981**
C_0	122.4	124.1
C_3 low	77.8	82.5
C_5 low	65.8	72.9
C_3 high	162.2	171.4
C_5 high	202.2	176.0
Kentucky 31	91.0	102.0
Missouri 96	111.8	131.2
LSD (5%)	27.9	23.1
CV (%)	15.8	12.7

From Reeder, L. R., Sleper, D. A., and Nelson, C. J., *Crop Sci.*, 24, 97, 1984. With permission.

western part of the U.S., while the primary area of consumption is located in the tall fescue belt described earlier. Significant genotype × environmental interactions do occur for seed yield.[25] As a result, it is necessary to evaluate breeding materials simultaneously for seed yield in the northwest and for herbage productivity in the area of consumption, which is often far removed from the area of seed production.

Because high seed yields are not always compatible with high herbage yields, the breeder must often decide to what extent seed yields can be sacrificed for higher herbage yields. The extent of this negative relationship between seed and herbage yield depends on the source of the breeding material. Not all breeding populations exhibit a strong negative association. Highest seed yield can be achieved most successfully by selection for early maturity and for those plants having high numbers of short panicles.

3. Resistance to Diseases and Insects

Most of the tall fescue hectarage in the U.S. is infected with an endophyte known as *Acremonium coenophialum*. When cattle graze tall fescue infected with the endophyte, they develop symptoms such as a rough hair coat, elevated body temperature, poor weight gain, poor milk production, and decreased feed intake. These symptoms are referred to collectively as summer syndrome or summer slump.[26] Hoveland et al.[3] reported on a 4-year study that showed steers grazing infected tall fescue had average daily gains of 0.50 kg, while those grazing noninfected tall fescue (less than 5% infection) had average daily gains of 0.83 kg.

The presence of the endophyte has important implications in the breeding of tall fescue cultivars.[27] When developing forage cultivars, breeders are interested in eliminating the endophyte because of the problems associated with animal performance. The endophyte infecting tall fescue appears to be seed transmitted only.[28] The level of infection in a particular pasture is related to the percentage of the viable endophyte-infected seeds used to plant the field.[29] Fields planted to endophyte-free cultivars have remained free for many years.[30] The endophyte can be removed from seed by aging, fungicides, or heat treatment. The most convenient means of eliminating the fungus is to age the seed under temperatures of between 25 to 30°C for approximately 1 year. Higher temperatures and humidity are more damaging to the viability of the fungus than are cooler and drier conditions.[31]

Tall fescue breeders developing turf cultivars wish to include the endophyte because of its reported association with resistance to several insect pests. For example, Barker et al.[32] have demonstrated a reduction in feeding and oviposition by the Argentine stem weevil (*Listronotus bonariensis*) on tall fescue and perennial ryegrass infected with the endophytic fungus. In the U.S., perennial ryegrass turf cultivars containing the endophyte have greater resistance to sod webworms (*Crambus* spp.) and bluegrass billbugs (*Sphenpphorus parvulus* Gyllenhal).[33,34]

The incorporation of the endophyte into desirable plant materials can easily be accomplished by making crosses to infected plants and using the infected plants as females. Because the endophyte is transmitted only by seed and not through the pollen or other known means, outcrossing to infected female plants represented the most convenient procedure of developing infected, improved turf cultivars. The injection of the endophyte into selected genotypes via a hypodermic syringe has been largely unsuccessful on a large scale required in a turf breeding program. Tall fescue turf breeders are at the forefront of using the endophyte as a means of breeding for enhanced resistance to insect pests.

Most of the diseases affecting tall fescue are incited by fungi.[35] One example of fungus that causes disease in tall fescue is *Drechslera sorokiniana* (Sacc.) Subram. & Jain (= *Helminthosporum satimum* P. B. & K.). Linscombe et al.[36] discovered that the inheritance for resistance to this pathogen was largely additive. There is little information on viral diseases. Tall fescue is known to be a host for barley yellow dwarf virus. Nematodes have also been reported associated with tall fescue.

Few tall fescue breeders have a primary objective of breeding for resistance to insect pests. Little information is available on how important insects are in damaging tall fescue swards. A survey was conducted in Missouri to evaluate the occurrence of leafhoppers feeding on tall fescue.[37] The investigators identified 54 species from 37 genera in 8 subfamilies. *Agallia constricta* Van Duzee, *Draeculacephala antica* Walker, *Endria inimica* Say, *Exitianus exitiosus* Uhler, *Forcipata loca* DeLong & Caldwell, *Latalus sayi* Fitch, *Polyamia weedi* Van Duzee, and *Psammotettix lividellus* Zetterstedt represented 74% of the total adult and 88% of the total immature populations suspected of having economic importance in tall fescue. Another insect identified feeding on tall fescue was *Forcipata loca*.[38,39] Damage from this pest was more severe in reducing forage quality than herbage yield. Studies cited above were conducted without knowledge of the level of infection of the endophyte described earlier in this chapter. It is not known what affect the presence of the endophyte has on the feeding habits of the insects just mentioned.

4. Herbage Quality

Tall fescue breeders are putting concentrated efforts into development of cultivars with improved herbage quality. Many different objectives are currently being researched in this area. These include breeding for increased *in vitro* dry matter disappearance (IVDMD), increased palatability, lower cell wall contents, higher Mg concentrations, and higher protein levels, just to name a few. Because tall fescue forage cultivars will be consumed largely by ruminants, it is important to improve herbage quality with the ultimate goal of improving animal performance.

Narrow-sense heritability for IVDMD has been reported to range from 27 to 66%.[40] In the same study, broad-sense heritability estimates were zero for IVDMD because no genetic variation among parents was found for any harvest date. It appears that more information on heritability obtained from a broader range of genotypes is necessary to obtain an accurate assessment of the genetic variation available for IVDMD.

IVDMD data collected from several hundred clones of tall fescue in the fall showed a wide range in IVDMD values.[8] The average IVDMD was 64.7%, and the range for the several hundred genotypes was from 55 to 74% (Figure 1). This indicated that variation for IVDMD may be sufficient to select for improved IVDMD in tall fescue, particularly during the fall. Because of the wide range in IVDMD, a selection program was initiated in the fall to select for higher IVDMD. Five cycles of selection have been completed at the Missouri Agricultural Experiment Station, and experiments are presently being conducted to measure what progress, if any, has occurred in selecting for higher IVDMD.

One of the most convenient means of selecting for improved IVDMD is to select for late-maturing plants. As maturity advances, IVDMD decreases. The biggest deterrent to this approach, however, is that late-maturing cultivars often have lower seed yields.

Hypomagnesemia, or grass tetany, is a disease that ruminants succumb to when utilizable Mg is not adequate. Magnesium may not be available to ruminants because concentrations are low in the herbage, or because interaction of other substances within the plant or animal lead to lower blood serum Mg levels.[41] Kemp[42] reported that no cases of clinical grass tetany occurred at blood serum Mg levels above 0.19%. A proper balance between

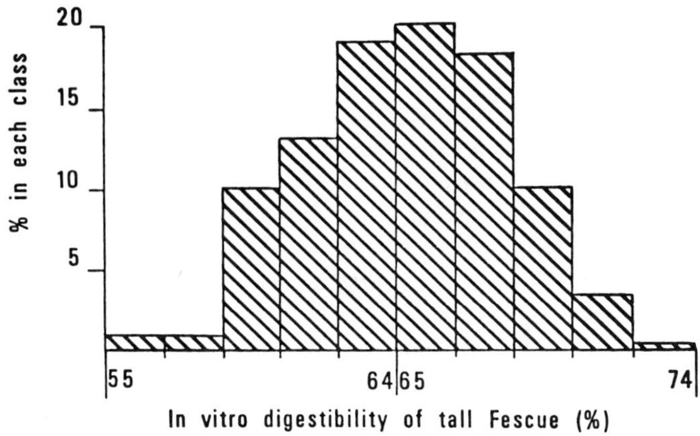

FIGURE 1. *In vitro* digestibility of tall fescue genotypes evaluated in the fall by rumen fermentation. (From Sleper, D. A., *J. Plant Breed. Rev.*, 3, 313, 1985. With permission.)

three mineral elements, namely, Mg, K, and Ca, also appears necessary to reduce the incidence of grass tetany. Kemp and t'Hart[43] and Butler[44] concluded that with a K/(Ca + Mg) ratio of less than 2.2 meq/100 g dry matter, the incidence of grass tetany was reduced. Grass tetany has been observed most frequently in ruminants grazing lush, cool-season forage grasses, including tall fescue, in the spring or fall.[45]

Narrow-sense heritability estimates for Mg accumulation in tall fescue is high, with values ranging from 85 to 91% with low estimates reported for the cation ratio of K/(Ca + Mg).[46] Selecting for high levels of Mg in tall fescue herbage appears to be a worthy objective.[47]

The whole area of developing cultivars with improved herbage quality takes on added significance because the poor animal performance that was observed with tall fescue in the past has been largely attributed to the presence of the endophyte. With the elimination of the endophyte, it is now possible to realize further improvements in animal performance as the result of developing cultivars with improved herbage quality.

C. Steps in Cultivar Development

1. Source Nursery

The first step in development of tall fescue cultivars is to collect germplasm that will serve as good sources from which to initiate selection. Germplasm that is obtained will be established in the field and placed into large source nurseries. It is not uncommon for source nurseries to include 10,000 or more individual plant genotypes. Entries in source nurseries may be replicated depending on resources and the supply of seed.

Because tall fescue did not evolve in the U.S., much of the germplasm collected for inclusion in the source nursery will be from overseas. The U.S. Plant Introduction System maintains seed stocks of tall fescue collected from the sources of origin. Other sources of germplasm for inclusion in the source nursery may include old cultivars, natural ecotypes or land races, germplasm pools created to produce unique gene combinations, and specific crosses among selected parents, to name a few.

2. *Phenotypic Selection*

Plants included in the source nursery will be examined on a phenotypic basis. The plant breeder establishes objectives for a new cultivar and looks for desirable characters in the plants included in the source nursery. Characters of interest, for example, might include high IVDMD, high herbage and seed yields, increased leafiness, increased tillering, persistence, and disease resistance. Evaluation for the selected characteristics may continue for more than 1 year because tall fescue is a perennial species and repeatability of performance is often required. After data have been recorded on plants in the source nursery for a specified period of time, desirable genotypes are chosen based on phenotypic performance.

Only a few plants out of the entire source nursery will be selected. It is not uncommon for a breeder to select only 200 or less plants out of an initial source nursery that may have included 10,000 or more individual genotypes. Seeds are harvested from desirable maternal plants from the source nursery and may be composited to form a new breeding population for further selection or to use in progeny testing.

3. *Progeny Testing*

Because tall fescue cultivars are marketed as synthetics, it is necessary to identify parents of putative experimental synthetics that have high general combining ability. General combining ability is defined as the average performance of a strain in a series of crosses.[13] General combining ability is associated with high additive gene action. Only those parents that have high general combining ability will be intermated to produce an experimental synthetic. Breeding procedures that are often used by tall fescue breeders to identify superior high-combining-ability parents include either the topcross, polycross, or the open-pollinated progeny test.[15]

4. *Experimental Synthetics*

Genotypes that have been identified as having high general combining ability are isolated in a crossing block to produce the first synthetic generation seed (syn-1) for testing. The number of parents used to produce a tall fescue synthetic cultivar ranges from 2 to as high as 30 or more. Most breeders use a high number of parents to minimize the effects of potential inbreeding when the seed is advanced for several generations. To obtain sufficient seed for sale to producers, it is often necessary to advance the seed for four or more generations.

The syn-1 generation is produced under conditions of random mating. This is most often accomplished by establishing the selected genotypes in a polycross nursery. The polycross nursery has each genotype replicated enough times so that all genotypes have an equal chance to mate with all individuals in the crossing block. For example, if a breeder develops an experimental synthetic which includes ten genotypes, conditions of random mating could be approached by replicating each clone ten times in the crossing block. The breeder may choose to make pollinations by hand to improve upon the chances of having random mating. Equal amounts of pure live seed is bulked from all ten replicates for each entry in the polycross nursery to obtain syn-1 seed. Successive synthetic generations are advanced through open pollination.

5. Small Plot Trials

Next, the experimental synthetics are established and evaluated in small plots along with check cultivars. Plots are usually established at several locations and evaluated from 3 to 5 years. Data collected from small plots will depend upon the objectives that the breeder established for the development of a new cultivar.

It is necessary to advance the seed to the syn-2 or later generation before inclusion into small plot trials. There is often a great deal of heterosis expressed particularly in the syn-1 generation, and as a result, the performance of the syn-1 generation may have little resemblance to the later generations which producers will purchase.

After completion of the small plot evaluation phase, the most desirable one or two experimental synthetics are chosen for inclusion in animal trials. It may also be desirable, if sufficient seed is available, to establish strip tests in farmers' fields to see how the new synthetic(s) performs on a larger scale.

6. Animal Evaluation

Animal evaluation is the last phase of cultivar development prior to release. Breeders have often tried to obtain animal evaluation of breeding materials earlier in the selection process. This has not been feasible because sufficient herbage is usually not available to conduct meaningful animal trials during early phases of cultivar development. The number of entries one can test with animals is certainly limited because of the resources required in conducting a high-quality evaluation. Techniques used in the evaluation of tall fescue experimental cultivars by animals will likely be related to output per animal and output of animal product per hectare.[48]

7. Release

The development of a new tall fescue cultivar takes approximately 10 years. After completion of the evaluation phases, the cultivar needs to be advanced for several generations so that sufficient seed is available to meet demands of the producer. During the seed multiplication phases of release, care needs to be taken so that the new cultivar breeds true for the characteristics for which it was selected. Production of breeder seed (syn-1) is under the direction of the plant breeder. The foundation class of seed is produced by the institution or company that developed the cultivar. The registered and certified classes are produced by selected growers, who in turn sell to producers.

D. Sources of Genetic Variability

Because tall fescue is included in a polypoid complex that ranges from 2x to 10x, there are ample opportunities to improve upon the genetic variability through interspecific and intervarietal hybridization (Table 1). However, most tall fescue breeders have not taken advantage of this approach to date. Considerable enhancement of genetic variability has occurred through intergeneric hybridization with *Lolium*. As mentioned earlier, the cultivar Kenhy was developed through this approach.[12]

Tall fescue is cross-fertilized, and a high degree of self-sterility exists. As a result, tall fescue is highly heterozygous and leads to heterogeneous populations. Tall fescue breeders have relied largely on the existing genetic variation in *F. arundinacea* as source materials to make progress through selection. Sources of genetic variation relied upon include planned crosses among selected clones, commercial cultivars, and populations improved by recurrent selection.

There is a need to improve upon the genetic variation of tall fescue by increasing the number of accessions from the source of origin. The Plant Introduction System in the U.S. has only several hundred accessions of tall fescue in its collection. This is in sharp contrast to the cereal grains, where it is common to find many thousands of accessions per species. This is not enough to satisfy future demands for tall fescue cultivar development. Tall fescue breeders need to be concerned that the center of origin for the species may lose the array of genetic variation because many natural habitats have been or will be destroyed or altered by human encroachment.

E. Breeding Procedures

Breeding procedures used to develop cultivars of tall fescue are those that can accommodate additive gene action. Because hybrid tall fescue cultivars have not yet been developed, breeding procedures that accommodate nonadditive gene action are not used. The four most popular breeding procedures used to develop new tall fescue cultivars include recurrent phenotypic selection, topcross test, polycross test, and the open-pollinated progeny test.

1. Recurrent Phenotypic Selection

Recurrent phenotypic selection is based on the phenotype of individual plants. A type of recurrent phenotypic selection procedure, often referred to as mass selection, involves selecting desirable plants after open pollination has occurred, harvesting the open-pollinated seed, and bulking it to obtain the population for the next cycle of selection. Selection is based on the maternal parent only because no control exists over the source of the pollen parent. Breeders can improve upon the efficiency of recurrent phenotypic selection by imposing maternal and paternal control by isolating selected male and female plants for the purpose of producing seed for the next cycle of selection.[23] Recurrent phenotypic selection has its greatest utility in the early phases of a breeding program, when little or no previous selection has been done. Breeders often go into source nurseries and select desirable phenotypes and combine harvested seed from these selected plants without progeny testing. It is a rather inexpensive method to develop improved populations from which other breeding procedures could be used to develop improved cultivars.

2. *Topcross Test*

Desirable genotypes are identified in the source nursery and outcrossed to a broad-based tester. The resulting progeny are evaluated in a progeny test that may consist of several years of small plot trials replicated at two or more locations. After completion of the small plot trials, parental genotypes are selected that have high general combining ability, based on the performance of the progeny test, and later combined to form experimental synthetics.

When using the topcross test in identifying parents with high general combining ability, care needs to be taken in choosing the broad-based tester. The tester needs to contain a broad genetic background and be widely adapted so that it can be useful in identifying genotypes that will likely intermate and produce progeny that have superior performance over a wide range of environmental conditions. A good choice for a broad-based tester, for example, might be a population that was improved by phenotypic recurrent selection or a broad-based cultivar such as Kentucky 31.

3. *Polycross Test*

The polycross breeding procedure is often used after previous phenotypic selection has occurred.[15] The improved population resulting from recurrent selection is (1) established at one or more locations, where random mating occurs among the entries, (2) harvesting the open-pollinated polycross seed of each entry separately, (3) and using the seed to evaluate each entry in a replicated polycross test.

The first step is to establish a population of several hundred selected genotypes that have been cloned through tillers. Clones are space planted at random and planted in as small of an area as possible to minimize environmental influences, such a soil heterogeneity. The several hundred clones are phenotypically evaluated for traits considered important by the plant breeder Out of these several hundred clones, 25 to 50 are selected for inclusion in a polycross nursery.

The polycross nursery is established with clonal replicated ramets to insure the production of seed under conditions of random mating. To further improve upon the chances of having random mating, clones must be of similar maturity, have similar plant height at maturity, have uniform volume of pollen shed, have uniform cross-compatibilities, and have no self-fertilization and lodging.

Seed is harvested separately from each plant in the polycross nursery. An equal amount of seed is bulked from all replicates of each clonal entry in the polycross nursery. This seed is then used in a polycross progeny test. The polycross progeny test is grown at one or more locations for 2 to 4 years with two to four replications at each location. Polycross progenies with the best performance for desirable traits determined by the breeder are identified.

Superior clones identified by the progeny test as having superior general combining ability are used to form experimental synthetics. These experimental synthetics are then evaluated further for their potential release as improved cultivars.

4. Open-Pollinated Progeny Test

The open-pollinated progeny test is one of the simplest and least expensive breeding procedures.[15] It involves selecting desirable genotypes from a large population, such as a source nursery, and harvesting seed from plants having superior traits of interest. Selection is based entirely on the female plant, with no regard given to the male parent. The selected plants most likely will be pollinated by nearby plants with similar anthesis dates.

Next, the open-pollinated seed is used to plant a replicated progeny test at one or more locations for several years, similar to that used in the polycross breeding procedure. The progeny test is used to identify those clones with superior general combining ability. After the superior parental clones are identified, they are intermated to form experimental synthetic cultivars. These experimental synthetics are evaluated for their potential release as a new cultivar.

IV. POTENTIAL USE OF HAPLOIDS AND DOUBLED HAPLOIDS IN A TALL FESCUE BREEDING PROGRAM

As plant scientists working with a highly self-infertile, wind-pollinated perennial, we must consider advantages and disadvantages of various approaches to genetic improvement of this important species. Basically, the goal is to eliminate undesirable genes and to incorporate superior genetic characteristics into stable breeding lines and/or cultivars.

New technologies, whether used alone or in conjunction with conventional procedures, should be considered if they can accelerate accomplishment of this goal. During recent years, many optimistic statements have been made concerning the potential use of new technologies such as gene identification and chromosome splicing, DNA transfer, induced mutation and selection at the cellular level, somaclonal variation, gamete-derived haploids, etc. Regardless of how the genetic code is manipulated, the critical factor is regeneration of fertile plants that are genetically stable and superior to existing genotypes under real-world conditions.

From a plant breeder's viewpoint, it seems most realistic to start with material that already has a high probability of cell-to-cell genetic variability, to regenerate plants from such cells, to evaluate the regenerated plants under field conditions, and then to use the most desirable ones alone or in recombination with others. To date, gamete-derived haploids and their doubled haploids appear to be the most promising of the new technologies for potential use in tall fescue improvement. Tall fescue is ideally suited for this approach because the species is wind-pollinated and highly self-infertile, which results in a very high probability that each gamete is genetically different from all other gametes on the same plant and from other plants in the same field. Theoretically, culture of haploid plants from this vast array of gametic genotypes should then allow identification of superior and elimination of inferior haploid plants, followed by regeneration of fertile doubled haploid plants from the selected haploids.

Although it is still extremely difficult to culture large numbers of haploid plants from immature gametes of tall fescue, some important steps have been made by Kasperbauer et al.[49] The team used a new procedure that involved special media, timing, attached nurse tissue, and anthers from vigorous field-grown plants. The same group also developed a tissue culture method to induce endomitotic divisions in haploid somatic cells followed by regeneration of fertile doubled haploid plants from those cells.[50] Continued emphasis on refinement of procedures should be emphasized. The following discussion concerns potential use of new technologies in a tall fescue improvement program as viewed by a plant breeder.

A. Contrasts with Conventional Breeding Procedures

To date, tall fescue breeders have not used haploids or other biotechnologically derived genotypes in the development of superior cultivars. This is primarily because the production of haploids in tall fescue is still extremely difficult. Although Kasperbauer et al.[49] have successfully produced androgenetic haploids of this important species using an innovative nurse tissue procedure, more refinement is still needed for large-scale adaptation to a breeding program. When haploids of tall fescue can be produced more easily, they will undoubtedly provide opportunities to develop improved lines for incorporation into new cultivars.

With conventional plant breeding of tall fescue, little concern is given to how readily a given population can be effectively utilized, and in many cases modified, using classical plant breeding procedures, such as recurrent selection, for example. Most tall fescue populations readily lend themselves to conventional open pollination and to artificially controlled pollinations. Existing populations also lend themselves, for example, to clonal propagation via tillering, which is useful in multiplying genotypes in crossing blocks to increase the amount of seed produced. Virtually any population that the tall fescue breeder chooses to work with can be used without difficulty in all steps of the conventional breeding procedures discussed above. This may not be true when using haploids because early experience with cell and tissue culture approaches with crop plants has shown genotypic variability in the efficiency of producing functional plants.

Donor plant genotype, growth environment, age, and general vigor all affect the ability of immature pollen to develop into haploid plants. In *Nicotiana tabacum*, for example, there are wide differences in haploid production among cultivars, and even among different plants within the same cultivar.[51] If this same phenomenon exists in tall fescue, it increases the importance of developing and using donor plant growth environments and anther culture media and environments that offer the greatest probability of obtaining haploid plants. Some factors that may influence the frequency of haploid production include physiological state of the donor plants, developmental stage of the pollen, culture media, physical factors of the culture, cold pretreatment of the anthers, and pollen dimorphism.[52] For cultivar development via androgenetic haploidy, a high frequency of haploid production is desirable. Present evidence suggests that some populations of tall fescue may be more suitable than others for improvement via haploidy, screening to identify the best and eliminate the least desirable haploids, and development and use of doubled haploids from the selected haploid plants.

B. Inbred Lines for Hybrids

One of the major reasons why plant breeders of many crops are interested in gamete-derived haploids is because they have the gametic genetic code, and the doubling of haploids represents a rapid method of obtaining homozygous lines which are analogous to highly inbred, ''pure'' lines. The production and use of highly inbred lines has been successful in genetic improvement of many self- and cross-pollinated species. In cross-pollinated species, for example, the use of haploids and doubled haploids represents a potential means of reducing the time to obtain inbred lines for the development of hybrids, while for the self-pollinated species it potentially allows plant breeders to develop homozygous cultivars more rapidly than by conventional self-fertilization.

At the present time, tall fescue breeders find it difficult to develop inbred lines by conventional methods. Most tall fescue strains are highly self-sterile and, therefore, do not lend themselves to extensive self-fertilization to obtain inbreds. Because of the highly self-sterile nature of tall fescue, inbred lines have not been available. Therefore, haploids and doubled haploids represent potentially valuable plant breeding tools for the identification and elimination of undesirable recessive genes and for the production of homozygous inbreds free of the undesirable genes for the production of hybrids and/or recombination into new synthetic cultivars. Overcoming the barrier of self-sterility through use of doubled haploids as inbred lines for hybrid cultivars has great potential and can take advantage of the heterogeneity already available in tall fescue. Theoretically, elimination of undesirable genotypes during evaluation and selection of the best haploid plants and development of their fertile doubled haploids should result in substantial heterosis in hybrids between two doubled haploids. Heterosis has been reported for conventional tall fescue[53] breeding approaches, and the production of hybrids from selected doubled haploids offers a potentially novel means of producing superior hybrid cultivars, or synthetics involving recombination of more than two of the selected doubled haploid lines.

C. Use in Genetic Analysis to Study Inheritance of Quantitative Traits

Most of the important agronomic traits that the tall fescue breeder selects for are controlled by many genes. If effective selection programs are to be designed, the breeder needs to be aware of the type of gene action controlling a particular trait. The use of gamete-derived haploids and doubled haploids in tall fescue has the potential of facilitating and enhancing the understanding of quantitative genetics in tall fescue. For example, Choo et al.[54] suggested that haploids and doubled haploids can be used to detect gene interaction, estimate genetic variances, detect linkage, estimate number of genes, locate polygenes, and verify established genetic theories and hypotheses. Further quantitative genetic studies in these areas using haploids and doubled haploids should be of great benefit to the tall fescue breeder who desires to maximize expression of certain desirable plant characteristics.

Twin seedling-derived haploids within the *Festuca* sp. have been used to determine the effects of gene dosage on important physiological traits. A decaploid ($2n = 10x = 70$) genotype of *F. arundinacea* var. *letourneuxiana* was reported to have apparent photosynthesis that was significantly higher than hexaploid *F. arundinacea* Schreb. genotypes.[55] To determine if gene dosage or ploidy per se was influencing the rate of apparent photosynthesis, polyhaploids of the decaploid were obtained from polyembryonic seeds.[8] Analysis of the polyhaploids for apparent photosynthetic carbon fixation showed that this trait was not influenced completely by polyploidy per se because apparent photosynthesis of the polyhaploids was similar to that of the decaploid parent.[56] It is well known that additive gene action strongly influences the level of apparent photosynthesis.[57] However, it is hypothesized that natural selection had likely occurred for increased photosynthetic activity of progenitors before extensive polyploidization had developed to the decaploid level and that further effects of polyploidy on increasing apparent photosynthesis were minimal.

D. Use with Mutagenesis

In monoploid plants, all loci are present in the hemizygous condition. Monoploids represent convenient plant materials for the observation of recent mutable events. Because tall fescue is considered to be an allopolyploid, haploids from this species may not be as convenient for mutation studies as are diploids of other species. There are likely to be genes from the three distinct genomes of tall fescue that are in common, and this may prohibit the expression of certain desirable recessive mutations in polyhaploids. However, a major advantage for development and use of gamete-derived haploid plants from a self-infertile, wind-pollinated species such as tall fescue is that there is a high probability that each haploid plant differs genetically from haploid plants derived from other gametes from the same plant. Thus, the use of polyhaploids from tall fescue has the potential to facilitate procedures analogous to mutation breeding by providing plant materials in which many of the loci would be in the hemizygous condition. This should allow potential quantum leaps relative to conventional procedures.

Induced mutations have the potential of broadening the gene pool of tall fescue as they do in any plant species. The amount of time and effort needed to develop useful mutants can be enormous. Perhaps selection efficiency could be enhanced by inducing mutants of polyhaploid tall fescue cells and subsequently inducing these mutant cells to develop into whole plants. A potentially useful system for mutagenesis will likely be polyhaploid cell cultures, including protoplasts, callus, and microspores. Of course, induced mutations may result in both desirable and undesirable changes. Developing and screening mutants at the polyhaploid cellular level has potential in plant breeding only if cellular selections can be expressed at the whole-plant level; and traits such as maturity date, plant height, seed yield, and herbage yield could not be selected for at the cellular level. Screening mutants at the polyhaploid cellular level in tall fescue will likely be most successful for metabolic traits, herbicide resistance, and pest resistance. However, the mutated cells are useless in a plant breeding program unless whole plants can be developed from the mutated cells. Consequently, culture of haploid plants from heterozygous tall fescue plants, followed by cloning them through tillers and evaluation under a variety of field and stress environments, promises to be a much more realistic approach. This is especially true if chromosome numbers can be doubled in the selected haploids to produce fertile doubled haploid plants with the desired characteristics.

E. Use in Wide Crosses

Interspecific and intergeneric hybrids are possible in tall fescue. Tall fescue can hybridize with related species listed in Table 1, and hybridization with plants from the *Lolium* genera have been reported.[5,9] One could hypothesize that the transfer of genes using wide crosses could be facilitated by using haploids and doubled haploids. For example, in wide crosses including interspecific, intergeneric, and ecogeographic crosses, chromosomes can partially pair during meiosis. If haploids were cultured from these F_1 hybrids, some homozygous lines with heterozygous translocation chromosomes might be produced.[52] Using this technique, genes for improved forage quality, insect resistance, and disease resistance might be transferred from related species, genera, and ecogeographic races. In addition, it is possible that somaclonal variation among plants regenerated from somatic tissue cultures of wide crosses could be a useful source of genetic variation for the tall fescue breeder.

F. Use of Somaclonal Variation

Somaclonal variation has been observed for many important crop species.[58] Essentially, this refers to variation among plants that are regenerated from somatic cells derived from the same original plant. Somaclonal variation in plants is not completely understood, but some of the possible causes include gross karyotypic changes, cryptic chromosomal rearrangements, somatic crossing over and sister chromatid exchange, and transposable elements.[58] A combination of somaclonal variation, somatic cell fusion, cellular selection, and accelerated development of haploid and doubled haploid lines may provide useful techniques for inclusion in a tall fescue breeding program. Certainly, all new biotechnological approaches should be considered as potential tools, but it will continue to be important to keep the main focus on improved genetic combinations that are genetically stable and produce a superior crop.

V. POTENTIAL ADVANTAGES AND DISADVANTAGES OF HAPLOIDS IN BREEDING TALL FESCUE

The use of haploids in tall fescue cultivar development has not occurred to date. However, when androgenetic haploids and their doubled haploids are produced on a routine basis, they will provide a potentially useful tool to accelerate development of improved cultivars of tall fescue. Some important potential advantages of haploids include

1. Doubled haploids have the potential of being uniform for many important agronomic traits. At the present time, synthetic cultivars of forage tall fescue are characterized as being highly heterozygous and heterogeneous. Therefore, the use of doubled haploids to serve as parents for potential hybrids of tall fescue could lead to hybrids being uniform for such agronomic traits as maturity, herbage quality, and seed production. Uniformity of hybrid cultivars resulting from homozygous parents has the potential to facilitate the harvesting and grazing of a forage that would be uniformly high yielding and of high herbage quality.
2. If a reliable source of male sterility is found that could be used to produce hybrid cultivars, doubled haploids have the potential of rapidly developing homozygous inbred lines for use in hybrid combinations.
3. Selection among the homogeneous progeny of doubled haploids can be more efficient than selection among and within heterogeneous progeny

obtained during conventional inbreeding. Conventional inbreeding approaches are not useful in breeding tall fescue because of the high frequency of self-incompatibility. Thus, the use of haploids and doubled haploids has the potential of overcoming the barriers of self-sterility for the production of homozygous lines.

Disadvantages include

1. The production of haploids and doubled haploids requires some specialized equipment and skilled personnel. However, well-coordinated cooperation between the breeder and the skilled biotechnologist should prove highly rewarding.
2. Doubled haploid lines are expected to exhibit severe inbreeding depression. Doubled haploids will have many loci in a homozygous condition. This could be potentially deleterious by having less than satisfactory performance of important agronomic traits just as any other highly inbred plant. Breeders should incorporate doubled haploids into an outcrossing scheme in their plant improvement program to take advantage of the expected heterosis and to overcome potential adverse effects of inbreeding.
3. Frequency of haploids and doubled haploids from desirable breeding stocks may be low.
4. Preferential haploid production from certain genotypes may result in a nonrandom array of doubled haploids from a heterozygous source of gametes. Of course, the same is true for conventional procedures in which flowering of the desired genotypes may be out of synchronization.

VI. SUMMARY AND CONCLUSIONS

The culture of haploid plants from immature gametes represents a potential form of gamete selection for the tall fescue breeder. Gamete-derived polyhaploids are of great interest in tall fescue breeding because undesirable recessive genes could be identified and eliminated and because doubling of the polyhaploids should lead directly to homozygous lines. The identification and elimination of haploid plants with undesirable characteristics followed by production of homozygous doubled haploid lines from the selected haploid plants is likely to be the most useful aspect of using haploidy to improve tall fescue. It represents a potential means of eliminating undesirable and fixing desirable alleles. Tall fescue cannot readily be self-fertilized because of large amounts of self-sterility. Production of homozygous lines represents the potential for recombination of lines without undesirable recessive characters or for development of hybrid cultivars. Herbage and seed yield increases are desirable goals. The use of doubled haploids as inbreds in hybrid and synthetic cultivars represents a potential tool to further improve the agronomic performance of this important forage species.

REFERENCES

1. **Buckner, R. C., Powell, J. B., and Frakes, R. V.,** Historical Development, in *Tall Fescue,* Buckner, R. C. and Bush, L. P., Eds., American Society of Agronomy, Madison, WI, 1979, chap. 1.
2. **Spooner, A. E. and McGuire, W. S.,** Tall fescue pasture for growing and finishing animals, in *Tall Fescue,* Buckner, R. C. and Bush, L. P., Eds., American Society of Agronomy, Madison, WI, 1979, chap. 12.
3. **Hoveland, C. S., Schmidt, S. P., King, C. C., Jr., Odom, J. W., Clark, E. M., McGuire, J. A., Smith, L. A., Grimes, H. W., and Holliman, J. L.,** Steer performances and association of *Acremonium coenophialum* fungal endophyte on tall fescue pasture, *Agron. J.,* 75, 821, 1983.
4. **Borrill, M., Tyler, B., and Lloyd-Jones, M.,** Studies in *Festuca.* I. A chromosome atlas of Bovinae and Scariosae, *Cytologia,* 36, 1, 1971.
5. **Terrell, E. E.,** Taxonomy, morphology, and phylogeny, in *Tall Fescue,* Buckner, R. C. and Bush, L. P., Eds., American Society of Agronomy, Madison, WI, 1979, chap. 3.
6. **Evans, G. M., Asay, K. H., and Jenkins, R. G.,** Meiotic irregularities in hybrids between diverse genotypes of tall fescue (*Festuca arundinacea* Schreb.), *Crop Sci.,* 13, 373, 1973.
7. **Hunt, K. L. and Sleper, D. A.,** Fertility of hybrids between two geographic races of tall fescue, *Crop Sci.,* 21, 400, 1981.
8. **Sleper, D. A.,** Breeding tall fescue, *J. Plant Breed. Rev.,* 3, 313, 1985.
9. **Berg, C. C., Webster, G. T., and Jauhar, P. P.,** Cytogenetics and genetics, in *Tall Fescue,* Buckner, R. C. and Bush, L. P., Eds., American Society of Agronomy, Madison, WI, 1979, chap. 6.
10. **Chandrasekharan, P. and Thomas, H.,** Studies in *Festuca.* V. Cytogenetic relationships between species of *Bovinae* and *Scariosae, Z. Pflanzenzuecht.,* 65, 345, 1971.
11. **Chandrasekharan, P. and Thomas, H.,** Studies in *Festuca.* VI. Chromosome relationships between *Bovinae* and *Scariosae, Z. Pflanzenzuecht.,* 66, 76, 1971.
12. **Buckner, R. C.,** The tall fescue breeding program at the University of Kentucky, in *Proc. 30th Southern Pasture and Forage Crop Improvement Conf.,* University of Kentucky, Lexington, KY, 1973, 85.
13. **Allard, R. W.,** *Principles of plant breeding,* John Wiley & Sons, New York, 1966, chap. 21.
14. **Asay, K. H. and Sleper, D. A.,** Registration of Missouri-96 tall fescue, *Crop Sci.,* 19, 563, 1979.
15. **Sleper, D. A.,** Forage grasses, in *Principles of Cultivar Development, Crop Species,* Vol. 2, Fehr, W. R., Ed., Macmillan, New York, 1987, chap. 6.
16. **Nelson, C. J., Asay, K. H., and Sleper, D. A.,** Mechanism of canopy development of tall fescue genotypes, *Crop Sci.,* 17, 449, 1977.
17. **Zarrough, K. M., Nelson, C. J., and Coutts, J. H.,** Relationship between tillering and forage yield of tall fescue. I. Yield, *Crop Sci.,* 23, 333, 1983.
18. **Zarrough, K. M., Nelson, C. J., and Coutts, J. H.,** Relationship between tillering and forage yield of tall fescue. II. Pattern of tillering, *Crop Sci.,* 23, 338, 1983.

19. **Jones, R. J., Nelson, C. J., and Sleper, D. A.,** Seedling selection for morphological characters associated with yield of tall fescue, *Crop Sci.,* 19, 631, 1979.
20. **Sleper, D. A., Nelson, C. J., and Asay, K. H.,** Diallel and path coefficient analysis of tall fescue (*Festuca arundinacea*) regrowth under controlled conditions, *Can. J. Genet. Cytol.,* 19, 557, 1977.
21. **Horst, G. L., Nelson, C. J., and Asay, K. H.,** Relationship of leaf elongation to forage yield of tall fescue genotypes, *Crop Sci.,* 18, 715, 1978.
22. **Burton, G. W.,** Recurrent restricted phenotypic selection increases forage yields of Pensacola bahiagrass, *Crop Sci.,* 14, 831, 1974.
23. **Burton, G. W.,** Improved recurrent restricted phenotypic selection increases bahiagrass forage yields, *Crop Sci.,* 22, 1058, 1982.
24. **Reeder, L. R., Sleper, D. A., and Nelson, C. J.,** Response to selection for leaf area expansion rate of tall fescue, *Crop Sci.,* 24, 97, 1984.
25. **Nguyen, H. T. and Sleper, D. A.,** Genetic variability of seed yield and reproductive characters in tall fescue, *Crop Sci.,* 23, 621, 1983.
26. **Hemken, R. W., Boling, J. A., Bull, L. S., Hatton, R. H., Buckner, R. C., and Bush, L. P.,** Interaction of environmental temperature and anti-quality factors on the severity of summer fescue toxicosis, *J. Anim. Sci.,* 52, 710, 1981.
27. **Pedersen, J. F. and Sleper, D. A.,** Considerations in breeding endophyte-free tall fescue forage cultivars, *J. Prod. Agric.,* 1, 127, 1988.
28. **Bacon, C. W., Porter, J. K., Robbins, J. D., and Luttrell, E. S.,** *Epichloe typhina* from tall fescue grasses, *Appl. Environ. Microbiol.,* 34, 576, 1977.
29. **Siegel, M. R.,** Mode of transmission of the fescue endophyte, in *Proc. Tall Fescue Toxicosis Workshop,* University of Georgia, Atlanta, 1983, 48.
30. **Buckner, R. C.,** Genetic improvement of tall fescue and possible endophyte resistance, in *Proc. Tall Fescue Toxicosis Workshop,* University of Georgia, Atlanta, 1983, 59.
31. **Bacon, C. W. and Hinton, D.,** Biology of the endophyte of fescue: ultrastructural analysis and physiological relationships, in *Proc. Forage and Turfgrass Endophyte Workshop,* Oregon State University, Corvallis, 1983, 19.
32. **Barker, G. M., Pottinger, R. P., and Addison, P. J.,** Effect of tall fescue and ryegrass endophytes on Argentine stem weevil, in *Proc. N. Z. Weed Pest Control Conf.,* 36, 216, 1983.
33. **Funk, C. R., Halisky, P. M., Johnson, M. C., Siegel, M. R., Stewart, A. V., Ahmad, S., Hurley, R. H., and Harvey, I. C.,** An endophytic fungus and resistance to sod webworms in association in *Lolium perenne* L., *Bio/Technology,* 1, 189, 1983.
34. **Ahmad, S., Johnson-Cicalese, J. M., Dickson, W. K., and Funk, C. R.,** Endophyte-enhanced resistance in perennial ryegrass to the bluegrass billbug, *Sphenophorus parvulus, Entomol. Exp. Appl.,* 41, 3, 1986.
35. **Chapman, R. A.,** Diseases and nematodes, in *Tall Fescue,* Buckner, R. C. and Bush, L. P., Eds., American Society of Agronomy, Madison, WI, 1979, chap. 15.
36. **Linscombe, S. D., Watson, C. E., Jr., and Trevathan, L. E.,** Inheritance of resistance to infection by *Drechslera sorokiniana* in tall fescue, *Crop Sci.,* 23, 1085, 1983.
37. **Quisenberry, S. S., Yonke, T. R., and Huggans, J. L.,** Leafhoppers associated with mixed tall fescue pastures in Missouri *(Homoptera: Cicadellidae), J. Kans. Entomol. Soc.,* 52, 421, 1979.
38. **Quisenberry, S. and Yonke, T. R.,** Effects of *Forcipata loca* feeding on tissue of Kentucky-31 tall fescue, *Ann. Entomol. Soc. Am.,* 74, 521, 1981.
39. **Quisenberry, S., Macrae, T. C., and Yonke, T. R.,** Preference of *Forcipata loca (Homoptera:Cicadellidae)* adults to forage plants, *Environ. Entomol.,* 12, 1149, 1983.
40. **Nguyen, H. T., Sleper, D. A., and Matches, A. G.,** Inheritance of forage quality and its relationship to leaf tensile strength in tall fescue, *Crop, Sci.,* 22, 67, 1982.
41. **Mayland, H. F. and Grunes, D. L.,** Soil-climate-plant relationships in the etiology of grass tetany, in *Grass Tetany,* Rendig, V. V. and Grunes, D. L., Eds., American Society of Agronomy, Madison, WI, 1979, chap. 6.

42. **Kemp, A.,** Hypomagnesaemia in milking cows. The response of serum magnesium to alternations in herbage composition resulting from potash and nitrogen dressings on pasture, *Neth. J. Agric. Sci.,* 8, 281, 1960.
43. **Kemp, A. and t'Hart, M. L.,** Grass tetany in grazing milking cows, *Neth. J. Agric. Sci.,* 5, 4, 1957.
44. **Butler, E. J.,** The mineral element content of spring pasture in relation to the occurrence of grass tetany and hypomagnesaemia in dairy cows, *J. Agric. Sci.,* 60, 329, 1963.
45. **Grunes, D. L., Stout, P. R., and Brownwell, J. R.,** Grass tetany of ruminants, *Adv. Agron.,* 22, 331, 1970.
46. **Nguyen, H. T. and Sleper, D. A.,** Genetic variability of mineral concentrations in *Festuca arundinacea* Schreb., *Theor. Appl. Genet.,* 59, 57, 1981.
47. **Sleper, D. A., Vogel, K. P., Asay, K. H., and Mayland, H. F.,** Using plant breeding and genetics to overcome the incidence of grass tetany, *J. Anim. Sci.,* 1989, in press.
48. **Matches, A. G., Martz, F. A., Sleper, D. A., and Krysowaty, M. T.,** Selecting levels of herbage allowance to compare forages for animal performance, in *Forage Evaluation: Concepts and Techniques,* Wheeler, J. L. and Mochrie, R. D., Eds., American Forage and Grassland Council and CSIRO, Canberra, 1981, 331.
49. **Kasperbauer, M. J., Buckner, R. C., and Springer, W. D.,** Haploid plants by anther-panicle culture of tall fescue, *Crop Sci.,* 20, 103, 1980.
50. **Kasperbauer, M. J. and Eizenga, G. C.,** Tall fescue doubled haploids via tissue culture and plant regeneration, *Crop Sci.,* 25, 1091, 1985.
51. **Kasperbauer, M. J. and Collins, G. B.,** Anther-derived haploids in tobacco: evaluation of procedures, *Crop Sci.,* 14, 305, 1974.
52. **Chu, C. C.,** Haploids in plant improvement, in *Plant Improvement and Somatic Cell Genetics,* Vasil, I. K., Scowcroft, W. R., and Frey, K. J., Eds., Academic Press, New York, 1982, chap. 7.
53. **Moutray, J. B., Jr., and Frakes, R. V.,** Effects of genetic diversity on heterosis in tall fescue, *Crop Sci.,* 13, 1, 1973.
54. **Choo, T. M., Reinbergs, E., and Kasha, K. J.,** Use of haploids in breeding barley, *J. Plant Breed. Rev.,* 3, 219, 1985.
55. **Randall, D. D., Nelson, C. J., Sleper, D. A., Miles, C. D., Crane, C. F., Krueger, R. W., Wong, J. H. H., and Poskuta, J. W.,** Photosynthesis in allopolyploid *Festuca,* in *Nitrogen Fixation and CO_2 Metabolism,* Ludden, W. and Burris, J. E., Eds., Elsevier, Amsterdam., 1985, 409.
56. **Poskuta, J. W., Nelson, C. J., and Sleper, D. A.,** CO_2 exchange, leaf area and herbage yield of tetraploid and polyhaploid *Festuca* spp., *Photosynthetica,* 22, 341, 1989.
57. **Asay, K. H., Nelson, C. J., and Horst, G. L.,** Genetic variability for new photosynthesis in tall fescue, *Crop Sci.,* 14, 571, 1974.
58. **Scowcroft, W. R. and Larkin, P. J.,** Somaclonal variation: A new option for plant improvement, in *Plant Improvement and Somatic Cell Genetics,* Vasil, I. K., Scowcroft, W. R., and Frey, K. J., Eds. Academic Press, New York, 1982, chap. 8.

INDEX

A

B

C

F

G

H

S

T

U

V